Ubiquitous Positioning

For a listing of recent titles in the *Artech House GNSS Library*,
turn to the back of this book.

Ubiquitous Positioning

Robin Mannings

ARTECH
HOUSE
BOSTON | LONDON
artechhouse.com

Library of Congress Cataloging-in-Publication Data
A catalog record for this book is available from the Library of Congress.

British Library Cataloguing in Publication Data
A catalogue record for this book is available from the British Library.

ISBN 13: 978-1-59693-103-9
ISBN 10: 1-59693-103-5

Cover design by Yekaterina Ratner

685 Canton Street
Norwood, MA 02062

10 9 8 7 6 5 4 3 2 1

To my wife, Diana
and children, Frances, and Peter

Contents

Acknowledgements

Much appreciation is extended to the many people who have inspired and helped me with comments and suggestions about the content of this book. My apologies to anyone I have inadvertently missed in the list below. I would also like to thank my employer, BT, and the Computing Dept. of Lancaster University for the opportunity to study for a Professional Doctorate. This book is based on my literature review and the early chapters of my thesis.

John Abbott
Paul Bowman
Dave Brown
Jon Bryant
Peter Cochrane
Barry Crabtree
Ian Drury
Stephen Furner
Alan Hagan
Mike Hazas
David Hutchison
Lesley Gavin
Richard Gedge
Joe Finney
Dave Heatley
Ian Neild
Natalie Newman
Karen Martin
Chris Parker
Dipak Patel
Roger Payne
Ian Pearson
Tim Pinnock
Andy Radburn
Andrew Scott
Geoff Scott
Steve Wright

Robin Mannings, Martlesham Heath, U.K. March 2008

Chapter 1

Introduction

1.1 The Concept of Ubiquitous Positioning

This book is concerned with knowing accurately the position of things and people everywhere and what then follows from such knowledge. It is a common requirement in many areas of business and social activity and is being made easier by many recent advances in radio, computing, and the Internet. Many drivers now have satellite navigation units (or sat-nav) in their vehicles to avoid getting lost. People at work or at home are now using the Web to view maps and aerial photographs for all sorts of reasons. Geography and navigation of the outdoors has never been so accessible, but there are, however, many unmet needs, particularly if we are interested in inside the built environments and about how things change with time.

The current geographical aspects of the Web, sometimes referred to as the Geoweb, are largely static but in the real world where people and objects are often moving, we may need to consider real-time positional information. For example, when we look at a map we usually see a two-dimensional graphical representation but progress is being made in computerized 3D visualizations of streets and buildings. Since the real world has three dimensions of space, it is desirable to make more advanced graphics to use perspective and render the 2D map as a 3D view so people do not need to learn how to read maps. If changes with respect to time need to be included within a visualization, then we are really using a four-dimensional map that could show where specific vehicles, people, animals, or anything else that is important are situated in real time, within a scene that looks realistic. To deliver this sort of experience, it is necessary to use a wide variety of sensing technologies to detect where everything is, and how it is moving. Communications technology is then needed to deliver the information to those who need to use it, and behind the scenes, clever machine intelligence to automate and to make things simple for people by delivering views of the real world that are at least as good as the views computer games deliver in the virtual world.

"Whereness" will be used throughout this book as a convenient proxy for the rather verbose terminology "ubiquitous positioning." It has been coined by the author and many of his colleagues to encapsulate everything about accurate ubiquitous positioning technology, business, and their consequences. This book starts with today's technologies but also explains about the future. Today people use the satellite-based Global Positioning System (GPS) in sat-navs, and they have access to many existing maps and images that have been repackaged by various Web organizations. In the future, however, there will be many new radio, sensor, mapping, and computing technologies that will be used together. To help understand what we mean by Whereness, it might be helpful to consider a simple scenario.

For example, if I am meeting some people in an unfamiliar venue, although I could find the place on a Web map that corresponds to the postal address I have been given, it may not be obvious which actual building to enter, where to enter the building, where to go once inside, and so on. Could I check on the progress of my colleagues or know if the meeting is not viable due to some unexpected transport or security emergency? If my map-reading abilities are poor, then could I hope to see where to go as a real-world view similar to that generated in a computer game? If my vision is poor, then could some other media be used involving sound or touch?

From this simple scenario, we can see that for positioning to be ubiquitous (that is available absolutely everywhere), it must include working inside the built environment. The accuracy must be adequate to show unambiguously where people are located within a campus, a building, a room, as well as outside in the street, in the country, in a vehicle traveling, underground, or indeed, anywhere else. As well as location, the more general concept of position needs to be considered. Position includes orientation and is sometimes important. For example, is the person I want to meet heading towards me or in a direction away from me? Is my elderly parent suddenly horizontal because of a fall?

The position of objects as well as people is also sometimes important, especially if they are very valuable and keeping track of objects' positions can help prevent crime, improve safety, and manage logistics efficiently. Ubiquitous positioning will have an important part to play in managing a world that is getting less predictable as the climate changes and can also be a key element in the management of scarce resources and providing greater security in the face of new global threats. To give a flavor of the potential of Whereness, a number of scenarios and applications of its importance to the future will be discussed briefly, all of which will be explained in more depth later.

Obesity is a crisis facing the developed world. Whereness can help improve health and well-being as personal exercise will be monitored based on the actual physical movements people are making day to day. Footsteps and climbing will be detected and perhaps coaching offered to encourage people to walk and climb stairs rather than to use mechanical travel.

Rapidly aging population demographics are predicted in some regions, so the cost of care will soar. Whereness can help by monitoring the movements of elderly, young, and any other vulnerable people, to give them more personal freedom while offering a high level of care. Automatic tracking will alert care-givers if anyone is in an unexpected, undesirable position or location. These people-tracking services extend to animals and other segments of society such as offenders, who already benefit from more freedom, while authorities ensure security is tight but with a reduced workforce.

As climate changes and unpredictable weather becomes more common, there will be an increased need to sense and monitor the environment and to help emergency services. Very high-quality maps and logistics will be essential and automatic messages will need to be sent to those in danger. Whereness also has an important role in managing scarce resources. Transport can become more intelligent with better quality information, guidance, and automatic payment systems.

Although the issue of "Big Brother" is generally thought of as an undesirable aspect of the future, if the threat from terrorism grows we may have to learn to live with it. Tagging and tracking the innocent (rather than the guilty) could help greatly in the identification of suspects and the monitoring of unusual physical movements, particularly in crowded areas.

The negative aspects of Big Brother could be offset, at least to an extent, by more positive aspects of Whereness used to help deliver fun, games, social experiences, and culture. Whole new generations of new sports and games are becoming possible. As new computing diversifies, it is liberating the computer environment away from the traditional desktop into other environments. Computer games can, for example, be played on sports grounds and the recent social networking phenomenon will enter the real world of clubs, pubs, and the great outdoors generally. New forms of locative[1] media will provide new tools for conceptual artists to enrich culture and will give new media and performance new qualities.

To sum up, many important areas of life will be affected by Whereness in the future. Just as the Internet and mobile phone were disruptive technologies in the past, and the interactive Web, known as Web 2.0, is disruptive today, so ubiquitous positioning may be a significant disrupter of the future. Indeed, some would say it already is!

---

[1] Locative media concerns the delivery of computer multimedia, such as graphics, video, and sound, in a specific location.

1.2 The Aim of This Book

This book has two objectives: first, to inform a mainly business audience about what could be large economic opportunity and potential disruptive technology, and second, to help existing and aspiring technologists with the diverse nature of the relevant emerging and converging technologies. The author believes that neither technology nor business can be treated in isolation and hopes that both aspects of the book will be useful to both audiences.

Technologists tend to be specialized, but one of the characteristics of Whereness, and many other emerging opportunities, is that it requires a working knowledge of several areas. Whereness includes radio propagation, radio systems, sensors, and sensor networks, geographic information systems (GIS), mobile computing, Internet communications, and Web technology. Convergence is a recurring theme and many useful references are given so that the key areas can be followed up in more detail if required. Highlighting the importance and opportunities of convergence is a major objective.

The businesses associated with these technologies are relevant, but there are also new and growing movements concerning open systems. These challenge traditional commercial activities and have new commercial models that involve, for example, payments by advertising or no payments at all since they are "folksonomies" where people are sharing information for the common good. It is hoped that their significance will be appreciated.

By the end of this book, it is hoped that the reader will be convinced that Whereness cannot be ignored by anyone who is serious about the future of information and communications technology (ICT). In the future, absolutely everywhere that people and important things are situated will be mapped in high detail and be visible in an easy-to-visualize way via the Web (which will be truly mobile and ubiquitous also). People and objects will be visible in near real time, either directly or indirectly, and most computer applications will feed off this rich temporal-spatial information to become aware of position and greatly enrich lives, manage scarce resources, and keep us safe.

1.3 The Structure of This Book

1.3.1 Background and Overview

If time is short, then the first two chapters will give the reader a good introduction to all the important facets of Whereness. Chapter 1 starts with some history, background, and terminology and then continues with some general principles. Chapter 2 is an overview of Whereness that follows the structure of later chapters. If the reader is reasonably well informed about navigation, radio, and ICT, then this chapter may be skipped.

1.3.2 Motivation, Business, and Applications

Chapters 3, 4, and 5 concern the motivation for Whereness and are aimed at providing a commercial view of why anyone would want to pay for it. Chapter 3 takes a broad view of all the commercially important issues, discussing Whereness within the digital networked economy. Chapter 4 focuses on specific applications relevant to current businesses, which are grouped into two parts. First, there are applications that are relevant to intelligent transportation systems (ITS), often referred to as transport telematics (or just telematics) and second, those relevant to more general location-based services (LBS).

Chapter 5 is more futuristic and starts with a discussion of societal changes and the important part Whereness could have in the future. A number of potential important new application areas are discussed including reference to climate change and its possible consequences.

1.3.3 Technologies of Whereness

The following chapters are more technical and concentrate on the technology building blocks that are needed for Whereness. Chapters 6 and 7 cover all the important methods that can be used to find position physically. Chapter 6 is concerned with radio technology, covering topics such as global navigational satellite systems (GNSS), which includes GPS, cellular radio, WiFi, and Radio Frequency Identification (RFID). Chapter 7 covers the nonradio techniques to find position. Some of these are wireless, such as optical and ultrasonic systems, and others are mechanical, including mobile inertial systems and active flooring.

Chapter 8 is where Web 2.0 is discussed in detail and is about maps, mapping, and geographic information systems (GIS). This chapter is the most exciting in terms of change and impact, but is consequently likely to be the one that becomes out of date first. A review of current Web 2.0 Whereness players is given.

1.3.4 Whereness and the Future

The concluding chapter continues with technology but it concerns the future and has a strong research flavor. First, the future of radio and its spectrum is discussed along with other sensing methods. Second, a standardized framework for information concerning ubiquitous positioning is proposed using Semantic Web[2] methodologies. Finally, ideas from mobile robotic research are discussed, where people (instead of robot vehicles) can automatically build maps and maintain maps wherever they go.

---

[2] The Semantic Web is a movement in computer science and machine learning. It is seeking to structure and standardize information on the Web, so that both people and machines may use it more effectively.

The final part of the conclusion paints a vision of what Whereness may be like in terms of devices and services and what businesses need to do to turn the vision into a reality.

1.3.5 The Epilogue

Several more technical topics are explained in the Epilogue that may be useful to people intending to create instances of Whereness technology. It can be ignored by the more general reader.

1.4 General Principles

1.4.1 Some History and Terminology

To introduce some of the fundamentals of positioning, it is useful to look at how navigation has developed in the past. Let us consider first what we mean by navigation. Navigation is associated closely with positioning. It implies the participation in a journey, from a starting position to another position that is somewhere else, with perhaps other places passed along the way. We call these "way points" and can thus consider a journey to be a set of locations or positions separated in both distance and time. Location can be specified in many ways but in general there can be a symbolic or language-based reference, such as a place name or alternatively, a physical reference such as a latitude, longitude, and height expressed in numerical form, derived perhaps from taking some measurements or from a map.

Navigation, historically, has had an enormous impact politically, socially, and economically as mariners learned to find their way out of sight of land. They used their eyes and simple instruments to observe the movements of the heavenly bodies (weather permitting) from which they could find latitude and their orientation and heading. The magnetic compass was used in the Middle Ages, at first in the form of a natural magnetic iron oxide know as lodestone. It provided a reliable heading in most places. Today we might describe these traditional navigation methods as sensing techniques and systems: collecting information by detecting (or sensing) the presence of physical phenomena.

Maps and their maritime equivalent, charts, were drawn, and as the understanding of mathematics, geometry, and instrumentation advanced, they developed from a generally artistic endeavor with largely symbolic information, into accurate graphical scale drawings. Cartographers eventually found solutions to the problems of representing the three-dimensional solid and nearly spherical Earth as a two-dimensional drawing. During the Renaissance when the voyages of discovery were in progress, they used projections using geometry to distort the image in various ways. One of the most well known is the Mercator Projection, which is unique in that it preserves direction so that bearings taken from a map can be used directly to navigate, but this came at the expense of distorting greatly the size and shape of anything near the poles. We are fortunate today to be able to use really simple interactive Web-based map tools that allow a much better visual understanding of geography.

Measuring the angle above the horizon of the sun at noon or a polar star at night gave an accurate reading for latitude but finding longitude was much more difficult. Originally, the only method was to measure progress by the technique of dead reckoning that involved estimating speed and direction. The traditional method was by timing and counting the passage of knots tied at regular distances in a line attached to a float thrown overboard. It remained nearly stationary and thus pulled the line along at the same speed as the vessel. As a voyage progressed, the accuracy of the estimate was diluted because errors accumulated. The fortunate navigator would be able to correct the positional estimate when the lookout spotted a way point that was recognizable. The estimate plotted on the map would be corrected and the process started afresh. Dead reckoning is still used as the basis of many navigational systems and has the advantage of not relying on any external measurements. The technique of map matching to correct positional estimates is also still in use in systems as cheap as road vehicle navigational devices and as expensive as cruise missiles.

A big step forward occurred with the invention of the chronometer [1], which allowed navigators to have constant accurate knowledge of time. The difference between noon, as measured from the time of the sun reaching its zenith (the highest point in the sky) and that shown on the chronometer gives the longitude (by the simple calculation that one hour [positive] difference is equivalent to 15 degrees of longitude [east]. The Greenwich Meridian in East London was established as the international datum of zero degrees (see Figure 1.1). By the age of the steamship, nonelectronic navigation had matured to the point where navigation was reasonably accurate provided the sky was visible often enough.

National mapping agencies were established. For example, the British Ordnance Survey started its work at the time of the Napoleonic wars at the end of the 18th century. National agencies used accurate observations, measurements, mathematics, and cartography to map everywhere of economic or military significance. Maps also included topography (i.e., were three-dimensional surfaces) with land heights and sea depths represented in various ways. Mapmaking is still very important, except it is now highly automated and includes aerial photography from aircraft, satellite imaging, and both radio and laser ranging.

The next revolution in navigation came with the invention of wireless radio communications. It was soon found that radio could be used in two very useful ways: first, as a method to pass information over very long distances and second, as a sensing system to detect and guide aircraft and ships. Radio and optical systems use electromagnetic waves that travel at light speed and are easily reflected off many surfaces, especially metals. Primary radar was developed in the late 1930s to sense the range, height, heading, and velocity of aircraft and ships by bouncing radio signals off remote targets and sensing the reflections. The time taken for a pulse to travel to a target and back again provides a distance measurement (known as lateration). The angle of the antenna (known as angulation) when it illuminates the target provides a heading and height. Speed is found by finding the Doppler shift in the received radio signal frequency that could be measured by mixing it (or heterodyning it) with the radio frequency carrier of the transmitted signal. Secondary radar was also used to determine the identity of the target and is active rather than passive. The first systems were military "identification friend or foe" (IFF) systems. A radio transponder fitted to the "friend" squarks [sic] a valid digital code when polled by the system (unlike a foe that would remain inactive). This signal would also be picked up by the highly directional primary radar antenna. The "blip" on the screen from the primary system could thus be labeled with its identity. Identity is of fundamental importance to positioning. Today we may see the same transponder approach being used by air traffic controllers and in shops and warehouses with goods fitted with small RFID tags.

Further advances in the Second World War (WW2) led to aircraft guidance systems that used a combination of radio timing measurements from different transmitters to provide an automatic means to find the location of exactly where to drop bombs at night. These electronic advances set the stage for the space age when electronic navigation matured and systems such as GPS became possible. Autonomous guidance systems or autopilots were developed for both aircraft and rockets that used gyroscopes to provide a stable orientation platform much more reliable than a compass. Accelerometers and ground topography following radar added further sophistication, accuracy, and automation.

The underwater equivalent of radar is sonar, which uses much the same approach but uses sound pressure waves in water rather than electromagnetic radiation. Indoor positioning is currently being researched and developed using ultrasonics in air. The main differences between sound and radio are that sound wave propagation is much slower and the range is less.

Passive radio techniques for positioning were also first used in WW2 and were generally known as direction finding (DF). When a radio signal was intercepted, an array of receiving antennae was used to create a highly directional system that had high antenna gain in one direction. This could be steered to find the bearing of the signal. If repeated in other places, then the intersecting beam directions could be plotted on a map and the approximate location of the transmitter found. A more automatic approach to this method involved the invention of the goniometer, where a static fixed circular array of receiving antennae could be made to "rotate" electronically by continuously switching the antennae to the receiver. Many modern radar and radarlike systems use static antenna arrays with complex computer controlled electronic beam forming techniques to create and steer beams rapidly without moving physically.

Before GPS was available, some dedicated terrestrial navigation services were developed, for example, the Loran system [2]. It uses networks of fixed transmitters that broadcast signals to mobile units. The technique is known as hyperbolic navigation because the position is found at the intersection of a number of hyperbolae, which result from the hyperbolic line of position (or loci) of places where the phase of the broadcast signals is equal.

Other commercial systems such as Quiktrak [3] were developed that work in a reverse way to that of Loran-like systems. They use mobile transmitting beacons (or sometimes transponders) and networks of receivers. The transmitter "bugs" can be very small and can be used to track animals and (stolen) vehicles. Some systems use networks of fixed receivers and others use mobile receivers to follow the targets.

The ability to place satellite radios in Earth orbit greatly advanced the scope of radio navigation. In a similar way to how ground-based systems developed, there are satellite systems that track moving transmitters (e.g., Qualcomm's OmniTRACS system) and others such as GPS that only broadcast reference signals.

GPS, which is owned and operated by the U.S. government for military purposes, is available free for civilian use. It offers highly accurate positions anywhere where there is a clear line of sight to at least 3 satellites (of the 24 that orbit slowly across the sky). The Soviet Union started to build its equivalent system, GLONASS, which is now being developed further by Russia. The EU has proposed Galileo, a global civilian commercial system, and the Chinese government is working on Beidou and Compass.

At about the same time as the satellite systems were being deployed in the 1980s, there was a revolution in mobile communications as cellular radio became widespread. The second generation of these mobile phones was digital (e.g., the GSM system), which made two-way mobile data ubiquitous. These systems were very useful for telematics applications involving GPS to communicate positions but also for positioning based on the use of the cellular network itself. Both cellular approaches have led to many location-based services.

Navigators do not need to understand how to perform the calculations involving time and space; it is all automated by computer software. The basics are all still there behind the scenes but the computer provides much more than the basic navigational calculations based on the measurements. It can give a symbolic position rendered in textual format, the spoken word, and all manner of graphical and pictorial formats. Until quite recently, advanced multimedia navigational computers were only available to the military and aerospace sectors, but ICT costs have fallen so much that that they are now available at any mass marketer.

Timing references are fundamental to Whereness. Mechanical chronometers were replaced as soon as electronic oscillators became possible, and a great step forward was the quartz crystal oscillator (with a maximum accuracy of around 1 part per billion). Even more accurate atomic clocks have subsequently been developed that are now used as international standards. Some are small enough to be carried in navigation satellites.

There has been a general trend where early adoption of advanced technology used by the military is reused by civilian mass markets. Whereness is highly dependent on this "peace dividend" and military technology still remains a fruitful place to find new technology for the consumer. Many of the WW2 techniques are being used again in ubiquitous computing experiments, which in due course will be commercialized.

1.4.2 Today's Opportunities and the Current Fragmented Market

The situation today is that there are many ways to find position but they are all separate and use technologies that differ. The raw information is used independently and computing applications do not have any way to access position as a general service. There are many useful applications but they are operating in "stovepipes" or vertical businesses seeking to address specific and usually niche markets. The emerging applications and market areas are currently fragmented. One of the problems with the whole idea of Whereness is that it concerns bringing together very different areas of specialist knowledge, some of which have been known for many decades (maps, navigation, and radio ranging) and some of which are quite new (digital wireless, the Internet, ubiquitous computing, and the Semantic Web).

The two most important areas are location-based services (LBS) and intelligent transportation systems (ITS), with ITS a subset of LBS. The main difference is that LBSs are mostly provided by mobile network operators and ITSs are associated with transport organizations and public authorities. It is the intention of this book to present a converged approach, taking the view that the spread of computing and communications will dominate the future and lead to mass markets replacing the niche areas. Whereness may become a hugely significant economic opportunity. The voyages of discovery and subsequent globalization were predicated on reliable navigation and positioning, but before discussing applications and the business aspects of Whereness it is important to consider a few basics.

1.4.3 Position, Location, and Coordinates

There is a difference between what we mean by location and position. Hightower's taxonomy[3] of location systems for mobile computing applications [4] draws the distinction that a position can include orientation. For example, a wheeled vehicle could be traveling backwards or forwards, or in the case of an aircraft that can travel in three dimensions, position could include information about yaw, pitch, and roll.

---

[3] A taxonomy of positioning (after Hightower) is included is in the Appendix.

We usually have some sort of reference frame, usually the center of the Earth, and thus can describe position with six parameters: three for location (latitude, longitude, and height above sea level) and three for orientation (yaw, pitch, and roll). In Figure 1.1, the position of a flying aircraft is fully specified. Although the location has an absolute reference frame, the orientation is likely to be within a relative reference frame, derived perhaps from the direction of travel (which would be relative to the absolute location reference frame). Location is, however, sometimes relative; for example, when specifying the position of one object with respect to another.

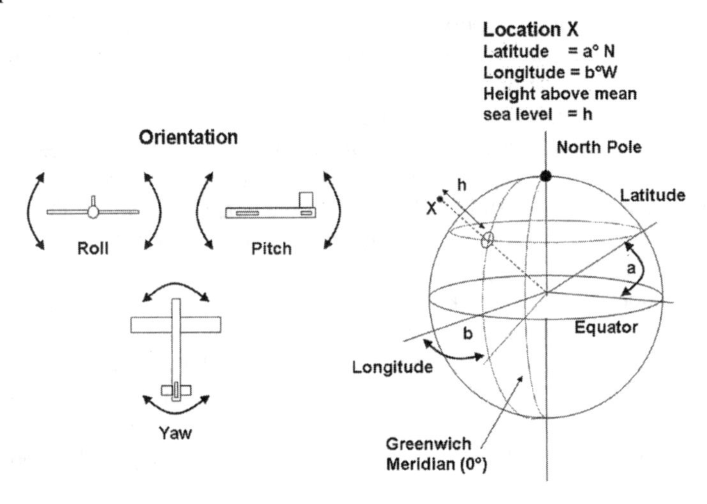

Figure 1.1 The three dimensions of location and three dimensions of orientation.

Maps usually use a grid system with Cartesian coordinates specified by "eastings" and "northings," rather than the polar coordinates of latitude, longitude, and height (usually above sea level). Simple algorithms can be used to convert between the different systems [5].

1.4.4 Remote and Autonomous Positioning

Most positioning systems can be categorized according to where the processing of the sensed information takes place. It can done locally inside the mobile equipment of the user, remotely at a fixed radio base station, or remotely at a computer center. Each approach has it merits. A remote system can minimize the investments in user equipment, but it places the onus of processing and detection centrally. As user numbers grow, the system does not scale well, since at some point it is likely that more central equipment investments and radio channels would be needed. In contrast, an autonomous system such as GPS scales very well (as far as the infrastructure is concerned). Once the satellite and ground station infrastructure is in place there is virtually no limit to the number of simultaneous users. The downside is that there is considerable complexity at the user end. The trend in distributed computing, however, is to push intelligence to the edge of networks so it is likely that the autonomous approach will dominate. Drane [3] makes the point that a hybrid approach is probably optimal.

1.4.5 The Inadequacies of GPS

Given the dominance of GPS it is a good place to start when considering Whereness. GPS was such an important milestone that it is now hard to imagine a world without it, but there are, however, some concerns and limitations. Perhaps the biggest concern is not about the technology per se, but about ownership and control of the service. Apart from the user groups for whom it is operated, there are no service-level agreements in place for anyone else. For an increasingly global economy with more people dependent on GPS in more countries, it is not surprising that the political control of GPS raises doubts about its value. One way around the problem has been proposals and deployments of GPS-like systems from other power blocks.

GPS is dependent on a network of fixed ground stations for monitoring and control. While it would be unhelpful for one of the 24 individual satellites to fail, it would be much more of a problem if an overall system failure occurred. Civilian aircraft may not use GPS as their only navigation device because of this vulnerability. Another good reason to have parallel deployment of systems such as GLONASS and Galileo is to provide system diversity (using multistandard receivers).

Although GPS works well in many environments outdoors, it has only limited coverage indoors. It would be useful if there could be methods to provide positioning of similar performance (i.e., better than 10m accuracy, where sight of the sky is not possible). Although conventional radio systems such as cellular radio and broadcasting services can all be used for positioning indoors, there is little prospect of achieving GPS-like accuracy or better. Given the small scales of indoor "geography" it would be useful to have better than GPS accuracy. So is there any prospect of solving the indoor positioning problem? Since the early 1990s ubiquitous computing researchers have been using techniques that, if adopted universally, would give an adequate performance indoors, and some are now beginning to become available in devices that are part of the ICT market.

1.4.6 Ubiquitous Computing

Whereness is dependent upon the future of ICT and the digital networked economy. We must therefore consider where the Internet, telecommunications, computing (both hardware and software), wireless, sensor and sensing systems, maps, and GIS are all headed. Although the desktop PC, mobile phone, and the portable music player are truly ubiquitous, there are (only) about 1 billion PCs globally and approximately 3 billion mobile phones. In the decades to come, the number of networked devices will rise to trillions if the predictions for ubiquitous computing are realized.

This movement in computer science is known by several names, including ubiquitous computing, pervasive computing, ambient intelligence and (in perhaps a less academic vein) as "The Internet of Things." Weiser [6], in a key paper in 1991, outlined a vision which, as devices have grown smaller and cheaper, is gradually beginning to happen. The first really large deployment of small ubiquitous computing devices is the take up of RFID technology by supply chain managers and ticketing agencies.

Most of the devices will have a computing capability and be internetworked using wireless communications, and many will have the ability to sense the environment. New human interfaces and software interfaces will emerge and a more decentralized and distributed approach to computing is also likely. Computing per se becomes a more abstract concept as the actual information processing is decoupled from any particular device. It could be considered to be a utility, available on demand from which every networked device may be available with capability and capacity. Ubiquitous computing has a "24/7/360°" quality where 24/7 implies that services are always available at any time and 360 degrees implies it is available all around, in any place [7]. Philips Research uses the term ambient intelligence and has presented its thinking in the book, The New Everyday View of Ambient Intelligence [8].

Some devices will be sophisticated and have much the same capability as a PC but be embedded within an electrically powered appliance that then becomes an Internet-appliance. Others will be extremely simple and be powered by scavenging power from the environment and only be on-line infrequently. In between these extremes will be devices of varying complexity. For example, at the low end of the market are a new generation of wireless security systems and domestic animal tags and at the high end, devices that stream wireless audio and video entertainment around buildings.

There is much more to ubiquitous computing than merely the increase in device numbers; however, it is the density of networked computing devices that is of interest to Whereness. The prospect is that in the medium term, virtually anywhere where people congregate there will be devices that can be used to provide accurate positional information. This is a fundamentally new possibility in positioning technology.

1.4.7 Context Aware Computing

User context is derived from the activities and situations of people and objects and is an important consideration in modern computing. Sensors and interfaces provide vital clues about the identity, attributes, and activities of the users who may be people or intelligent machines. Within a rich, fixed, and mobile computing environment, it is possible for the (distributed) machine intelligence to be aware of the context within which any computer applications are being used. Of all the contexts, the most fundamental are the identity of the subject, the temporal context (when events occur), and spatial context of position (where everything is located and its orientation) (see Figure 1.2).

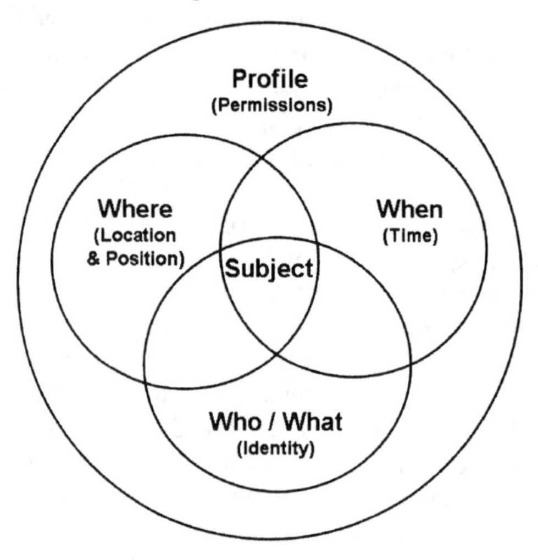

Figure 1.2 The fundamental contexts of who/what, where, and when.

1.4.8 User Profile

An important issue relating to user context is that of the preferences of the user; for example, whether the context of position is to be shared with any third party. Usually preferences are contained within a personal profile that can either be created manually or inferred. The profile can be used at both the server side and the user side of any session, but generally, it is hosted and backed up at a server operated by a service provider. Security and trust (arguably the most fundamental services being provided in any on-line session) are managed by reference to the profile and it is important to note that the inclusion of a Whereness context can greatly facilitate their management. Whereness context can thus be useful in two ways. First, it can be used to increase security and trust generally and second, to support specific LBS applications. Presence is another important issue handled by on-line services. It includes information such as "out at lunch." Although currently manually controlled, a Whereness service could perhaps automate many presence services.

1.4.9 A Priori Information

The Latin expression "a priori" is used to describe information that is known already. It is very important in context aware systems. Although some context can be inferred after measurements have been made, it seems reasonable to use what information might already be known in advance. Examples include digital maps, building plans (the indoor equivalent of maps), digital diary or calendar entries, swipe card data, and details of any wireless and sensor systems.

1.4.10 Positioning Practicalities

There are many ways to find position at one extreme using systems deployed in space and at the other extreme, using systems where range is zero (i.e., the device is physically connected by a cable). Some systems use radio while others use the sensing of other physical phenomena that may be wireless or rely on some mechanical activity. These systems will be discussed in later chapters, but it should be clear that there is a need to find ways to use all these systems together in a way that maximizes utility and minimizes investments.

It is not difficult from a purely technical perspective to achieve ubiquity in positioning given the developments in computing and sensing already highlighted; however, the problem will be to do it in a way that is economical and that leads to sustainable businesses.

1.4.11 The Ad Hoc Approach to Sharing Location

A simple and rather obvious way to find location is to ask someone close by who already knows. If machines are involved, then the positional information can be passed or exchanged as digital messages. This approach is at the heart of another major movement in ICT: the idea of peer-to-peer information sharing and the spreading of information in a "viral" fashion. By taking advantage of chance physical meetings, useful information can be passed on. The Latin term "ad hoc" is used to refer to unplanned informal activities and is used to describe short-range wireless sessions between devices that are positioned in an unplanned and informal way. Thus a person or thing that has good positional information could use an ad hoc wireless system to pass this information to someone who is close by. If they both have some positional information, then it is likely that the combined information will be better for both parties. To preserve trust, before any ad hoc system can operate, the user profiles have to allow such operations.

1.4.12 Quality of Information

A big challenge in positioning is the quality of processed information, which is dependent on the accuracy, precision, latency, and availability of underlying the raw data. It is likely that commercial opportunities in Whereness will come from the automatic management and increase in the quality of information.

Most technologies also have a dark side. No doubt a new sort of hacker might emerge with the objective to spoof false positions. Great care is therefore needed to take potential abuse into account. It would be unfortunate if rogue users are telling people false personal positions and informing others falsely of their own locations but even worse, if false maps were to be created deliberately.

One of the big business opportunities will be about the management of Whereness by highly trusted organizations. They will be trusted to manage profiles, to manage identity, to validate sensor information, to validate mapping, and to host and distribute all a priori information.

1.4.13 Robotic Systems and Machine Intelligence

It may not seem obvious why robotics is relevant to positioning but if we consider an autonomous machine that can move (i.e., a mobile robot), then knowing its position accurately is usually a fundamental requirement. A body of research is building based on the needs of mobile robotics that includes not only the ability to automatically follow a map but also to build the map in the first place. A classic experiment involves mechanical "mice" finding optimal routes through mazes. A more recent idea is substituting a human for the robot where the electronic sensory systems and guidance algorithms are interfaced to a person who can both build and then follow digital maps. Once a map has been built and validated it can then be shared with other users.

If Whereness is destined to be truly ubiquitous then an automatic approach to the building and updating of maps is necessary. The process of information automation is an aspect of machine intelligence or artificial intelligence (AI), another major movement in current computer science research. One aspect of AI is the development of the Semantic Web. It is about giving information more meaning so that it can be interpreted by both humans and machines. The long-term objective is to turn the Web into a massive database and one of the earliest domains to be tackled in this way is digital mapping.

1.5 Summary

In this chapter ubiquitous positioning or Whereness was defined: it concerns knowledge of accurate position of important objects and people in all environments and computer applications dependent on that knowledge. Geography delivered via the Web (known as the Geoweb) and GPS-based satellite navigation are touching the lives of increasing numbers of people in the outdoor environment, but this book outlines how the built environment and all other areas can be mapped and included. Maps are changing from 2D graphical images into 3D visualization (common in computer games) and Whereness is promising to bring a temporal aspect so that maps and visualizations are four dimensional and reflect the real-time aspects of the real world.

The book is aimed at both a business and a technical audience and has a number of objectives: first, to give enough information to appreciate all the important aspects of Whereness and to give references to allow readers to follow up in more depth; second, to stress the importance of convergence since no single technology or dependent business is likely to be adequate; and third, to highlight the evolving open approach to mapping and other relevant information. This includes Web 2.0. (a more collaborative Web model) and is being extensively used to deliver existing maps and images via Web services and also the more disruptive approach of "folksonomies," where information is shared freely for the common good. There is a real prospect that the open approach to mapping and positioning may map the world in great detail in all environments. A final objective is to convince readers of the importance of Whereness as part of the digital networked economy so that the vision can become a reality.

A brief history of navigation is given to introduce unfamiliar terminology and to give an early appreciation of the many topics that will be covered in later chapters. The rather fragmented current market for some early Whereness applications in location-based services and intelligent transportation systems is introduced and is followed by some general principles that include the following topics.

The difference between location and the more general position that includes orientation is explained. How scalability is enhanced by autonomous positioning rather than a more centralized approach. Why GPS is not adequate. The importance of ubiquitous computing, a current research theme, is highlighted together with some related concepts concerning a priori information, personal profiling, quality of service, the need for an economically pragmatic approach to infrastructure, and the relevance of mobile robotics.

The next chapter continues to introduce relevant topics and gives a complete overview of Whereness.

References

[1] Sobel, D., *Longitude,* New York: Walker, 1995.

[2] The International Loran Association, www.loran.org. Dec. 2007.

[3] Drane, C. R., and C. Rizos, *Positioning Systems in Intelligent Transportation,*Norwood, MA: Artech House, 1998.

[4] Hightower, J., and G. Borriello, "Location Systems for Ubiquitous Computing," *IEEE Computer Magazine*, Aug. 2001, pp. 57-66.

[5] Evenden, G., "Cartographic Projections Library Originally Written by Gerald Evenden, Then of the USGS," http://proj.maptools.org, Dec. 2007.

[6] Weiser, M., "The Computer for the Twenty-First Century," *Scientific American*, Vol. 256, No. 3, 1991, pp. 94-104.

[7] Buderi, R., "Computing Goes Everywhere," *MIT Technology Review*, Jan.-Feb. 2001.

[8] Aarts, E., and S. A. Marazo, *The New Everyday View of Ambient Intelligence*, Rotterdam: 010 Publishers, c2003.

Chapter 2

Overview of Whereness

This chapter details some further background to the areas of ICT relevant to Whereness and an overview of the main business application areas. An overview of positioning technology is given. This chapter and also the paper [1] written by the author are recommended for those who wish to grasp the gist of Whereness rapidly before getting immersed in the detail contained in later chapters.

2.1 Whereness in the Changing Digital Networked Economy

2.1.1 Convergence

Converging businesses are common in the area of ICT, for example, the offering by a single company of Internet services, fixed and mobile telephony, and TV. In contrast, some more traditional areas relevant to positioning such as public transport and motor industries tend to concentrate on their core businesses, the product cycles of which tend to be much longer. Convergence is about a coming together of activities that are separate and then diversification of the converged whole into new areas. New opportunities exploiting Whereness are therefore likely to be considered by organizations experiencing convergence in their traditional areas of business, especially mobile and fixed telecommunications operators, map content owners, and the electronic equipment market. Whereness applications would be regarded as valid adjacent markets and would promote the novel use of existing ICT products and services. It is likely that leveraging existing infrastructure investments will be more fruitful than building new dedicated infrastructures for Whereness services. The exception to this is the deployment of publicly funded systems such as GPS and Galileo. As internetworking spreads and the Internet becomes pervasive, the opportunities for Whereness are likely to increase as all the necessary parts are converged into efficient service offerings.

2.1.2 Disruptive Technology

Unexpected competition can come from business players outside existing industries either exploiting technology that is new or finding new ways to apply it. If the new value propositions are much stronger, then existing businesses can fail rapidly. The term "disruptive technology" was explored by Clayton Christensen in his seminal book, The Innovator's Dilemma [2].

Whereness may become a disruptive technology affecting existing niche players adversely. For example, companies offering small-scale vehicle tracking services may find their offering eclipsed by much more universal services from very large organizations. A wider set of applications would be offered covering entertainment, gaming, e-commerce and business-focused logistics applications; the economics of scale would apply.

The biggest current disruption is the impact of Web 2.0 and the ubiquity of services such as Google Maps and Google Earth, which now touch the general public in ways that other geographical systems have failed to do. New mobile devices such as the Apple iPhone are another potential disrupter. The availability of both potential disrupters within the same device is an exciting development that may change the face of both the mobile telephony and mapping businesses, together with other areas dependent on them.

2.1.3 Openness and Web 2.0

A recent trend in ICT has been a move away from the traditional closed or proprietary approach to products, content, and services to a situation where there is considerable openness. Openness can take several forms: first, having an entirely licensefree approach to standards, for example, the LINUX operating system and the Internet standards from W3C,[1] and second, creating a standard via an industry group, where anyone may join in but that involves paying a fee to join, for example, the mobile telephone standards. Whereness already includes aspects of both camps, but perhaps the most interesting trend is the third aspect of openness that concerns the opening of normally closed commercial systems via interfaces that anyone can use as a component part of a bigger system.

For example, although Google Earth is a proprietary system, it does have an open interface available through an application programming interface, or API. The API allows third parties to use underlying services but in novel ways that can create new applications. In fact it is possible to mashup several underlying systems via different APIs so that, for example, photographs hosted on a service such as Flickr can be combined with the maps from Google. This ability to share and combine (or mashup) is the defining aspect of Web 2.0 [3] and is of fundamental importance to Whereness (see Chapter 8).

---

[1] W3C is the standards body that standardizes Web technology.

2.1.4 Commoditization and Diversification

The digital networked economy (DNE) is supported by ICT, which continues to grow and develop. Its costs are falling and its problems relating to trust and security are getting fixed. Whereness, which is part of ICT, is therefore becoming more affordable and it can also help improve trust and security.

Web services are a cheap and convenient way to bring Whereness services to people without the need for special software. Businesses, for example, can geotag information on Web-based maps, and noncommercial information commons are also emerging where people share geographical information. As Whereness becomes widespread these systems will have dynamic content showing the current positions of moving (tagged) objects.

As established ICT players suffer increasing competition in their traditional offerings they will be motivated to move into new, more profitable areas such as digital rights management, trust services, and Whereness.

The scope of the Internet will widen as more devices and appliances become networked. This increases the scope for using networked devices to provide location references, in particular indoors. Wireless hotspots will increase in number and new radio systems will be used (e.g., WiMax and ZigBee).

The main business areas that can offer or exploit Whereness are:

- Global Navigational Satellite Services (GNSS);
- Cellular mobile radio operators;
- Wireless hotspot operators;
- Groupware providers;
- RFID[2] in supply chain management;
- Facilities management;
- Intelligent transportation systems (ITS);
- Video surveillance operators;
- Mapmakers and geographical information systems (GIS);
- Telcos, Internet service providers (ISP), and Contact Centers;
- Multimedia content owners and gaming operators.

Multiband satellite navigation chips will be included in many user devices and will be supplemented by cellular and hotspot positioning services. Corporate group software will extend the simple diary and calendar functions into real-time positioning services. Existing tag systems using RFID may become part of more general consumer and business ICT and play an important part in Whereness. Video surveillance and mobile camera technology can also extend its functionaly. Some ICT businesses will move into Whereness (as other more traditional areas

[2] Radio frequency identification (RFID) is explained in Chapter 6.

decline in profitability). Location-aware multimedia and gaming content will become important and extend the current business of content owners.

2.2 Today's Whereness Businesses

2.2.1 ITS and LBS

Intelligent transportation services (ITSs) in various guises have been developed since the 1980s to use ICT to make transport safer, more efficient, and better for the environment. They range from simple radio tagbased toll passes and ticketing systems, basic public information services broadcast by radio and more recently on the Internet, to much more sophisticated systems controlling city traffic and feeding automatic routes to individual drivers with navigation equipment. Almost all are dependent on positioning and location technology.

There have been attempts to base all services on a single system but the main conclusion after 20 years of research and development is that all the approaches remain fragmented. Part of the problem is that the main organizational players have widely differing objectives and timescales. Government and local authorities are slow to react, can be conservative regarding technology, and are accountable to the public. At the other extreme, the ICT and electronics industry is focused at providing equipment and services that are out of date sometimes in a matter of months, some of which are difficult to use and not always suitable for use in safety critical activities such as driving a vehicle.

One of the aims of this book is to suggest that ITS is only part of a much bigger set of opportunities. For example, transport applications normally focus on vehicles and passengers but the actual people involved usually only use vehicles for part of their journeys that normally starts and ends indoors. When he or she is traveling, the typical young passenger is likely to be using a personal music player and mobile phone (which now includes a camera), the business person a Blackberry (or whatever supersedes it), and anyone that is health conscious a pedometer-like device to count calories and monitor the body. All these are likely to have positioning technology included and all could be capable of delivering the main ITS applications.

Location-based services that are mostly associated with mobile phone applications have also been slow to develop. For reasons similar to why ITS applications have fallen short of their potential, LBS are probably also too fragmented to be really big businesses. Any service that for only works on a particular handset or from a specific service provider or that needs special skills to use is unlikely to be hugely popular. Perhaps the one exception would be if it is fashionable!

Once LBS along with ITS and all the other niche areas are treated as a whole, then the Whereness opportunity becomes greater and the investments needed to offer service become shared across more applications. It is important to remember that it is the location aspects of services that are delivering the value. A location-aware advertisement or traffic message is much less useful when it is not delivered at the right time, in the right place, in the right format, to the right person. This is difficult to do technically, and perversely, it can be an advantage to those who do. The organizations with the capability to converge all the systems necessary to deliver a good range of Whereness services are not likely to be many, and may have limited competition, at least for a time.

2.3 Future Whereness Applications

One of the problems with Whereness services to date has been the lack of a "killer application" so it is germane to ask if there will ever be one, or alternatively, will there be a set of niche applications as there are at present, which will then grow in significance?

There are such a diverse range of applications areas, some of which are so important to the future of society and business that the potential for killer applications remains, and Whereness can rightly be considered a potentially significant commercial opportunity. As an introduction to the application areas it is useful to consider first what are the potential new "hot" areas. Although more traditional areas of e-commerce remain opportunities (for example, location-based advertising), since the start of the new millennium there are certainly some growing trends that offer Whereness operators opportunities. Some of these will now be highlighted.

2.3.1 Management of Scarce Resources

The response to climate change will lead to pressure to reduce the greenhouse gas emissions from transport systems, which account for about a quarter of man-made emissions, and which may be achieved by many strategies—technological, economic, and moral. The effect will be a major change in the way vehicles are used so that the reason for every journey will be much more carefully considered. Whereas the original motivation for universal road tolling was to reduce congestion and raise tax revenues, in the future it may be much more about affecting behavior change to promote a lower carbon footprint. Organizations may well want to show their "green" credentials by keeping track of all business-related journeys and individuals may wish to keep a tally of personal emissions because they may gain tax incentives or perhaps a feel-good benefit.

Whereness services are clearly necessary to automate the process of keeping account of vehicle movements. It is not simply logging distances but also keeping account of acceleration and breaking, since the style of driving greatly affect the fuel economy.

Supply chain management is focused on the automatic collection of data about the movement of goods between field, factory, warehouse, supermarket, business, or residence. Today, tracking and tagging systems seek to increase efficiency by saving person-hours, reducing errors, and optimizing facilities. In the future it may be more about traceability, distances moved, and the reverse supply chain concerned with recycling.

2.3.2 Demography, Health, and Well-Being

Some would argue that the climate change issue started with demographic changes and that the problem is all about overpopulation now and its continued growth (from around 6 billion people now to perhaps 9 billion within a couple of generations). In the developed world, major problems are forecast concerning a rapidly aging population demographic. Managing a population with increasing impairments of mobility and ability will lead to increasing automation of care systems. One example will be to track the movements of those who need support and to raise alarms when the support is needed.

The major health and well-being issue of the future in the developed world is obesity. Today's generation is the first where the health of the "old" old will be better than that of the "new" old (where old is defined as over 50). Automatic systems to track personal movement and account for energy use are being researched. The simple approach is to use a pedometer to count footsteps; a more advanced system would monitor exact distances, the climbing of slopes and stairs, and using barcodes or tags to monitor food eaten. New sporting and gaming applications are also being developed that promote activity and include tracking and the delivery of multimedia based on location.

2.3.3 Self-Actualization

Since the industrial revolution, an increase in global wealth has had both positive and negative consequences. If overpopulation, climate change, and obesity are the unintended negative consequences often a positive consequence is more leisure and the resources to enjoy it. Figure 5.1 shows Maslow's famous hierarchy of needs. Much of the traditional business of Whereness is focused at the lower layers where the emphasis is on security and safety. Slightly higher would be the well-being applications and the systems to help with education (an example might be an automated information delivery service for use when walking around a historic site). At the top is "self-actualization," which includes things that engage with our emotions such as physical games and sports, which are both healthy from a physical perspective as discussed above and are also good from an emotional point of view, particularly when played with a group of friends. Collaborative tracking applications can mean that the friends can either be colocated or located anywhere.

Just as the social side of mobile phone messaging has led to many new social conventions and applications (and a very lucrative revenue stream for the operators), it is likely that the social side of Whereness will lead to new opportunities around the idea of meeting and coordinating with friends and colleagues and potential new contacts when out and about. This aspect of Whereness may well have the potential to dwarf the other applications and to be a truly killer application. Social engagement (in all its forms—many of which are illegal in many places) and gambling are the really big money spinners in the Internet world and adding a locational aspect to these is likely to be very popular with some people (however distasteful it may be to others!).

High-technology fashionable items are emotional choices. Dull and boring but perfectly usable technology is sometimes rejected and much more expensive "cool" devices are chosen. The motivation concerns image and style. It is possible that a range of location-aware jewelry or some other such style-related location-aware device may be very popular and hugely profitable.

Perhaps the ultimate in self-actualization concerns the future of art. Contemporary and conceptual art is blurring boundaries between what is defined as art, between artists and audiences (or spectators), and between the artistic media. The act of moving around is being used by conceptual artists to create works, where the participant's movements trigger the creation and delivery of music and visual displays that are part of living dynamic artworks in themselves. The artist has in effect created a machine that is creating the art.[3]

2.3.4 Humanizing Big Brother

An interesting side effect of using Whereness technology in a social, artistic, or gaming context may be to humanize it. One of the frequent negative comments made about tracking and positioning, especially when it involves people, concerns the Big Brother issue. Some issues of trust are due to misinformation, unfamiliarity, and difficulty in understanding what machines can do. Once people are familiar with the technology in a positive context they may be less afraid of it in others.

[3] "The idea becomes a machine that makes the art". - Sol LeWitt.

2.4 Radio Positioning

2.4.1 Communications and Sensing

Radio systems are fundamental to Whereness for two main reasons. First, radio provides a means to communicate position, and second, it can be used as a sensory system either with or without the communications functionality. For those systems that are intended for communications, the sensory aspects may be an added bonus. WiFi is a good example of a radio standard that is used for communications but is underused in its potential as a sensory positioning system.

2.4.2 Transparency

An advantage of positioning systems based on radio is that the signals propagate through many materials and environments with useful ranges that can be colossal. Most other approaches such as optical and acoustic sensing require an un-occluded line of sight between transmitter and receiver.

2.4.3 Far-Field and Near-Field Communications

Conventional radio frequency systems use the electromagnetic far field that has useful propagation ranges constrained only by the power available, the environment, antennae considerations, and regulations. In contrast, in the last two decades some very short range systems have been developed using near-field communication (NFC) for RFID tags. Ranges are typically from a few millimeters to 2 meters with the near field being used to transfer power as well as information.

2.4.4 Useful Radio Characteristics

Although radio characteristics such as frequency, modulation methods, and system architectures vary greatly, there are two basic methods to find location. The simpler involves measuring the strength of the signal, the power of which falls as the distance between transmitter and receiver increases, and its most basic form is simply the presence or absence of detectable signals. A more complicated and (potentially) more accurate approach measures the signal propagation times and angles.

Additional system elements may also be helpful. For example, in the GSM cellular radio system, the mobility management subsystem that allocates the radio resources contains location registers to keep account of the approximate location (by proximity) of the mobile stations. Known generally as "Cell ID," this information can be used for other purposes such as applications to detect fraud or to offer travel advice based on location.

2.4.5 Global Navigational Satellite Systems (GNSS) and GPS

By far the most popular radio positioning system is GPS. It is one of a number of existing and planned Global Navigational Satellite Systems (GNSS) and is provided by the U.S. Department of Defense, free for civilian use at reduced accuracy (about 10m) from that used by U.S. government agencies and their allies. GPS does suffer from a number of inherent problems, however. At least four satellites must have a clear line of sight to the receiver, which is not always possible in dense urban areas and inside buildings; it can be slow to synchronize so readings only follow power up after several minutes; and there are political considerations. Increased performance may be achieved by combining the basic GNSS with additional systems.

2.4.6 Cellular Positioning

There have been many proposals for enhancements to cellular systems to provide more accurate services than Cell ID. Most are based on the knowledge of signal timing, which is helped by the fact that digital radios send information in data packets that are inherently sequenced, highly organized, and time-dependent. It is unlikely that cellular positioning will ever be as accurate as GPS, but it does have the advantage of working indoors and in other areas where GPS signals are difficult to receive.

2.4.7 WiFi Positioning

WiFi positioning is a very useful and increasingly popular method. Most commercial WiFi positioning systems available today use simple proximity and are usually combined with communications. An interesting development is the amateur activity of wardriving, which is mapping WiFi access point location using GPS and providing the maps freely.

2.4.8 Ultrawideband Positioning

The most exciting new radio technology for positioning uses ultrawideband radio (UWB), which is classified as a spread spectrum system. It can be thought of as an indoor secondary radar system. Transponder tags transmit very lowpower nano-second duration pulses, which are inherently wideband in that they spread widely across the radio spectrum. Since the duration is so short, the timing accuracy can be very high leading to positioning accuracies in three dimensions of typically 10 cm.

2.5 NonRadio and Sensor-Based Positioning

2.5.1 Communications Cable Contact

There are many methods that can be used to find position and location that do not use radio waves and some could be considered rather obvious. First, the act of communicating by cable (i.e., using an electrical circuit or an optical fiber) can be used to show that the users are present at the termination. Telephone areas are a crude location system. A major problem now exists for emergency call centers since the new Internet Voice over IP (or VoIP) has no association with a line so cannot provide a location code. Clearly, it requires a Whereness service provided by some other means.

2.5.2 Electronic Diaries and Calendars

A second obvious location method is to rely on a diary or calendar entry. If someone has indicated that he or she plans to be somewhere then there is a chance that he or she will be there. If a wireless or sensor system can confirm the presence in some way, then the probability increases greatly.

2.5.3 Infrared

Some of the early experiments in indoor positioning used infrared (IR) communications. Since all optical systems have the disadvantage of needing a clear optical path between the transmitter and receiver, these experiments used electronic badges that had to be worn in a visible fashion. An advantage of these badges is the extreme cheapness and simplicity of using a IR light-emitting diode (LED) as the transmitter. Networks of fixed IR receivers can be fixed within rooms so that when a target enters the room, the code transmitted by the badge can be detected and a central database updated.

2.5.4 Ultrasonics

More advanced badge systems followed, substituting ultrasonics for the infrared devices. Although still needing a clear path, the use of sound has an advantage over any electromagnetic communications. The problem with wireless and optics is that signal propagation is at the speed of light ($300.10^6 ms^{-1}$), which makes timing measurements a serious technical challenge involving detecting nanosecond time periods. The speed of sound is six orders of magnitude slower and timings are thus well within the capability of the cheapest microprocessors. With ultrasonic systems centimetric positioning accuracy is both feasible and inexpensive. Range, however, is limited and to cover a building hundreds or even thousands of base stations would be necessary so infrastructure costs can be prohibitive.

2.5.5 Optical Video Cameras

Video surveillance cameras are useful to identify and track people and objects within view. Image recognition software is becoming increasingly powerful at dynamically extracting features from a video stream and following their movements within the scene. By using scene analysis it is thus possible to infer location. The approach is referred to as "outside in" video positioning. In contrast, "inside out" positioning uses mobile cameras looking out at scenes. For example, if a particular skyline or landscape was captured, it can then be compared with a stored library and then the location of the mobile camera can be determined. It is also possible to infer orientation. As mobile cameras are being incorporated into more portable devices, this approach is likely to become more important.

Cinematic applications use a form of video surveillance for motion capture with extremely high accuracy. Highly reflective dots may be attached to a target (perhaps an actor) whose exact positions can be recorded, digitized, and translated to support animation effects. This is perhaps the ultimate positioning technology in terms of accuracy but can only be made to work in a highly controlled and narrow-ranged environment.

2.5.6 Magnetic Fields

Oscillating magnetic fields have been used experimentally for indoor positioning. Graphic tablets use this technique, but if the "tablet" is made as large as a room, then it is possible to know the position of the sensing apparatus (usually an inductor coil) by the relative timing of the changing fields.

Magnetic compasses are also useful in mobile navigation systems since the Earth's magnetic field is almost always available except when distorted by ferrous metals and near the poles where there are little horizontal components. Thin film magnetic sensor technology is available cheaply so that solid, very robust state compasses are now common.

2.5.7 Mechanical and Inertial Systems

There are a wide range of mechanical systems that can be used to track people and objects. Active floors have pressure sensors that can monitor footfall as targets walk across them. Although some other system must be used to identify a target, it can then be tracked uniquely (assuming the sensor density is high enough).

Mechanical motion sensors are being integrated into portable electronic consumer equipment. Pendulum pedometers that are used for simple step counting are being replaced by more sophisticated and highly miniaturized solid-state chip accelerometer devices (using microelectromechanical systems, or MEMS) that are giving more and more devices the ability to detect motion in an increasing range of consumer devices, including the Nintendo Wii and the Apple iPhone.

Miniature solid-state gyroscopes are also available (used by some model aircraft autopilot systems) that together with the MEMS accelerometers give the possibility of high-quality inertial navigation at very low costs and sizes.

2.5.8 Sensor Fusion

Sensor fusion is a very important technique used to combine the information from different sensors. No sensing system is the perfect solution for all positioning requirements. If, however, a number of information streams from different sensing systems are "fused," then a greatly enhanced positioning capability can be gained. Integrating mapped information is also an important part of fusion.

In a typical system, where several sensors are providing physical measurements and a digital map is being used for map matching, there will be statistical algorithms taking the raw measurements and improving them.

2.5.9 A Summary of Positioning Methods

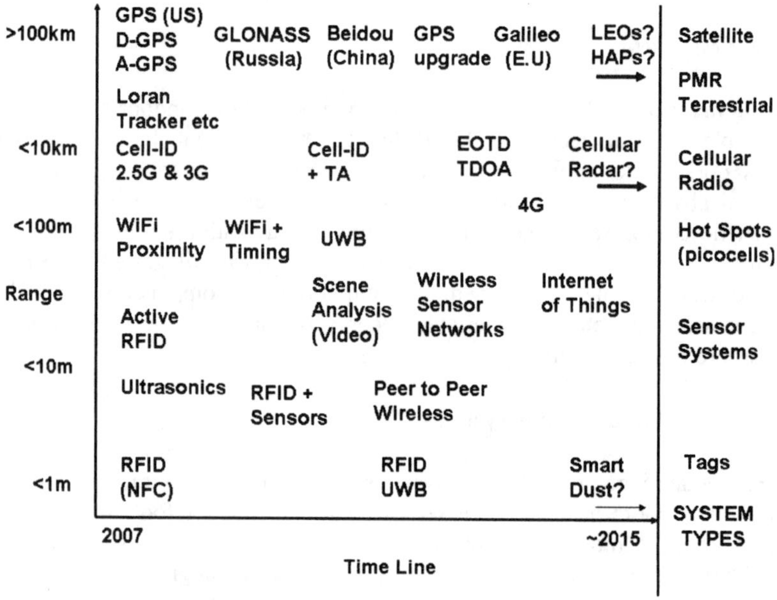

>100km
GPS (US)
D-GPS
A-GPS
GLONASS (Russia)
Beidou (China)
GPS upgrade
Galileo (E.U)
LEOs?
HAPs?
Satellite
PMR
Terrestrial
Loran
Tracker etc
<10km
Cell-ID
2.5G & 3G
Cell-ID
+ TA
EOTD
TDOA
Cellular
Radar?
Cellular
Radio
4G
<100m
WiFi
Proximity
WiFi +
Timing
UWB
Hot Spots
(picocells)
Range
Active
RFID
Scene
Analysis
(Video)
Wireless
Sensor
Networks
Internet
of Things
Sensor
Systems
<10m
Ultrasonics
RFID +
Sensors
Peer to Peer
Wireless
<1m
RFID
(NFC)
RFID
UWB
Smart
Dust?
Tags
2007
~2015
Time Line
SYSTEM
TYPES

Figure 2.1 Summary and time line for positioning technology.

Figure 2.1 shows all the principle positioning technologies plotted graphically. The horizontal axis is time and the vertical axis is approximate range, with families of technologies placed together horizontally as a series of time lines. Table 2.1 gives some details of these systems that are explained in more depth in Chapters 6 and 7.

Table 2.1

Summary of Positioning Technology

Type of System	*Name of System*	*Description of System*
Global Navigational Satellite System (GNSS)	Global Positioning System (GPS)	U.S. military and civilian (3m accuracy).
	Assisted GPS (AGPS)	Cellular radio data service enhances GPS performance especially for emergency calling (E911 services).
	Differential GPS (DGPS)	Ground stations generate positioning correction to enhance accuracy (<1m).
	Global Navigational Satellite System (GLONASS)	Russian system currently being upgraded.
	Beidou/Compass	Chinese system under development
	Galileo	EU System under development (<1m accuracy by 2013).
	Low Earth Orbit (LEO) Systems	Possible future disruptive technology (space tourism spin-off).
High Altitude Platform (HAP)		Possible future disruptive technology based on airships or solar powered aircraft.
Proprietary Systems	Loran - C (Long Range Navigation)	Long-established low-frequency maritime service (upgrade under consideration).
	"Tracker"	Example of vehicle antitheft system using private mobile radio (PMR).
	QuikTrak	Australian PMR system.

Cellular Systems	Cell ID (Cellular Identity)	Simple course positioning method using identity of cellular radio base station.
	GSM Localization: TA (Timing Advance)	Internal parameter used to align frame timing, which gives approximate range between base station and mobile.
	Time Difference Of Arrival (TDOA)	Centralized measurement of distance using signal timing from a mobile to at least three base stations.
	Enhanced Observed Time Difference (EOTD)	Similar to TDOA but with measurements performed locally at mobile (>50m accuracy).
	Fourth Generation Mobile (4G)	
	Cellular Radar	Experimental.
Hotspots or Picocells	WiFi (e.g., Ekahau)	By simple proximity (~100m accuracy). By timing (under development)
	Ultrawideband (UWB) (e.g., Ubisense)	Highly accurate indoor system (<10cm in three dimensions).
Sensor Systems	Active RFID (Radio Frequency Identification)	Battery-powered transponder (range <100m) positioning by simple proximity.
	Scene Analysis	Using cameras to recognize and position objects such as vehicle number plates (range depends on lighting and optics).
	Wireless Sensor networks (e.g., ZigBee [IEEE 802.15.4])	Used for control of buildings and environments. Positioning by simple proximity (<100m accuracy).
	"Internet of Things"	Experimental. Furniture and artifacts with embedded sensing and wireless (radio and infrared optical) widely deployed. Positioning by proximity (accuracy 10s m). Ad hoc approach to positioning possible. Includes use of inertial/mechanical sensing.
	Ultrasonics	Very accurate indoor positioning from transponders but clear line of sight needed (accuracy <10cm).

	RFID and Sensors	Battery-powered RFID transponders combined with sensing systems.
	Peer to Peer Wireless	Ad hoc approach to networks where mobile devices form local informal mobile networks. Nodes are all peers of each other with no master control system.
Tags	RFID using Near Field Communications (NFC)	Small printed tags powered by the near field become active only when in close proximity (<2m).
	RFID using UWB	Next generation RFID based on UWB.
	Smart Dust	Experimental concept where tags become similar in size to dust particles and are liberally deployed on surfaces powered by the environment.

2.6 Web 2.0 and Maps

Web 2.0 maps are becoming an increasingly important interface for many Web-based information resources. Chapter 8 details the various players that are creating digital maps and delivering them in novel ways. As Web 2.0 has developed, a number of very useful books have been published explaining the simple methods by which information can be mashed up by anyone with basic programming skills.

If positions are known (from any or several of the methods discussed above), then Web-based displays are the most convenient way to display positional information. Mapping underwent one revolution as mapmakers moved to digital maps and geographical information systems (GIS) that were used to create and manipulate them. Another revolution is under way with Web 2.0, but it is only one facet of the future.

Maps are moving from flat 2D entities into 3D representations where software models can be used to display landscapes in a very realistic format. In order to manipulate any digital image conveniently, the dot (or raster) format used originally[4] has been superseded by a more mathematical representation of shapes using vector graphics. It is not surprising that the same techniques that are used by computer games in virtual reality and in time "augmented reality" will result when the virtual worlds start to overlay the real world and augment it.

[4] Raster and vector mapping are explained in the Epilogue.

Adding time to maps adds a fourth dimension and dynamic mapping is very useful to manage events and visualize moving objects, for example, tracking vehicles and people. Maps can therefore be considered to be a multidimensional information space.

Open mapping concerns initiatives such as OpenStreetMap that are creating very high-quality maps using volunteers with simple GPS equipment to create log files of routes traversed. The maps are attributed and shared and represent a genuine disruptive technology.

2.7 Conclusion and Vision

2.7.1 Future Whereness Technology

Chapter 9 looks towards the longer term. Since positioning is heavily dependent on radio, a number of radio futures are discussed. First, moving to higher frequency bands where the atmosphere absorbs the radio signals that can be helpful to limit coverage and provide very high bandwidth picocells that will be useful for positioning. Second, the convergence of wireless standards is discussed, and finally cognitive radio. Positioning could play a key role in this new radio spectrum paradigm, since every radio could potentially know its position; then new ways can be exploited to configure radio links optimally.

2.7.2 The Semantic Web

Artificial intelligence is discussed with the use of the Semantic Web to help pull together the currently fragmented sources of information relevant to Whereness. An ontology for Whereness is proposed (which is an informational framework where everything is defined in terms of meaning), which would encompass existing geographical and sensor markup languages and extend definitions to include the radio environment and all other necessary classes of objects and property.

2.7.3 Simultaneous Location and Mapping

A final technical topic is simultaneous location and mapping (SLAM). Digital maps are very important to guidance systems as map matching can greatly increase the accuracy of positioning. However, if map matching is combined simultaneously with mapmaking, an automatic system can ensue that can create, correct, maintain, and, of course, follow digital maps. SLAM was developed from robotic research but is likely to be a key future element in Whereness since it promises to extend mapping indoors. Currently, digital navigation maps are very useful outdoors but do not continue into buildings and cannot deal with the concept of multiple floors within a building. The built environment is also more dynamic (especially in open plan offices where furniture is frequently moved) so a dynamic mapping approach is needed. Many research projects in ubiquitous computing are currently working in this area.

2.8 Summary

This chapter provided an overview of all the important business and technology aspects of Whereness. The importance of convergence was highlighted as a prerequisite. Whereness may be a genuine disruptive technology and be offered as a universal set of services by very large businesses already engaged in convergence. An important aspect to consider, however, is the threat or opportunity of open systems that is growing in importance as Web 2.0 makes progress.

A brief review of existing relevant businesses was given including intelligent transportation systems and location-based services. Future Whereness applications were discussed in the context of solving new challenges facing the world such as climate and demographic changes.

The technology of Whereness was introduced, including radio systems, sensor and sensing systems, maps, Web 2.0, the Semantic Web, and simultaneous location and mapping (SLAM).

References

[1] Mannings, R., "Whereness: Ubiquitous Positioning," *The Journal of the Telecommunications Network*, Vol. 4 Part 1, Jan.-Mar. 2005, pp.38-48.

[2] Christensen, C. M., *The Innovator's Dilemma*, New York: Harper Business, 2000.

[3] O'Reilly, T., *Web 2.0* ,http://www.oreillynet.com/pub/a/oreilly/tim/news/2005/09/30/what-is-web-20.html.

Chapter 3

Whereness in the Future Digital Networked Economy

This chapter concerns the motivation that leads to the provision and use of all the elements that are involved in the business of Whereness. Later chapters consider the technology in more depth, but here the perspective is customer and user wants and needs and the businesses that might evolve to fulfill them. There are a number of business models involved currently and there is reason to suppose there may be many more, since most involve ideas that are evolving with some still in the research phase.

In this chapter, the various business frameworks are discussed. The potential is considered for Whereness to create a major impact and perhaps be a disruptive technology, particularly as some business models may follow a more open and less obviouse commercial framework. The digital networked economy (DNE), the wider environment within which Whereness will exist, is described. More specifically, related businesses and systems are also listed, all of which will have a part to play. Whereness is fundamentally about convergence and integration.

The importance of the fundamental contexts of time, space, and identity is discussed together with the need for personal profiles to manage trust issues. Value-added services are often based on machine intelligence so four key intelligent functions are presented, which are routing information geographically, accounting values, route guidance, and controlling systems. The important issues of autonomy, quality of service, privacy, and trust are discussed because these are all part of successful service provision.

3.1 Convergence and Disruption

Technological convergence is now bringing together a number of domains that have been largely separate until now. These include fixed and mobile communications, geographical information systems (GIS), positioning, and sensing systems. Early adopters of positioning services are users, operators, and manufacturers of transportation systems. This is such an important area that several sections of the next chapter are devoted to intelligent transportation systems (ITS) of today and of the future.

It is likely that ubiquitous positioning may be a future disruptive technology and may have an impact on business and society that is far greater than may be commonly supposed. When the mobile phone was brought to market in the 1980s, nobody was forecasting that by 2003 there would be more mobile calls globally than fixed calls and that the market for mobile voice calls would dominate with now 2 billion handsets worldwide [1]. Perhaps a social "killerapp" for teenage users might emerge based on location-aware gaming or social group organization. This may develop in the same way as SMS texting, which was originally designed to promote voice calls for business people but led to a new social phenomenon, language, and culture, dominated by a youth market.

3.2 Commercial Frameworks

Is there any prospect for ubiquitous positioning to become a reality in the foreseeable future? There appear to be three ways for the provision of widespread Whereness services to become available.

First, it would be possible theoretically to build dedicated infrastructures to offer commercial services. The cost, however, would be of the same order as the provision of cellular radio (billions of euros or dollars per major network), and it seems unlikely currently that any single commercial system is likely to emerge or that the public would want or trust any government to do it. Second, although some dedicated systems are available [2], these are currently aimed at niche users and are likely to remain expensive options. The progress that has been made to date has been publicly funded with the GPS systems and the other regional GNSS funded by states. The third option is perhaps the most realistic scenario and draws upon the first two. This concerns the integration (or convergence) of many existing technologies and businesses to gradually extend positioning services with useful quality of service into all required environments and markets. It will, however, only become possible as the digital networked economy increases in its scope and impact.

3.3 The Impact of Openness

Users would no doubt like to have highly accurate positioning and location-dependent applications working everywhere and at all times, preferably for very little personal investment or even better, for free. Some of the business models associated with the Internet and computing take this approach and positioning technology is currently no exception. Is GPS really free? Are the advertisements enough to sustain all the Internet mapping services that are available currently? Will free open source software continue to be of growing economic significance?

In an economic analysis GPS would not be considered free since it is paid for out of taxation (directly by U.S. citizens and indirectly by the rest of the world who participate in the U.S. economy). The open source movement is supported by academia, which in most countries is supported by the public purse. However, the wiki movement [3] is a genuine "common" that shows no sign of "overgrazing."[1]

There is a strong likelihood that at least some elements of ubiquitous positioning may really be free. For example, if someone records the geography of a favorite walk, he or she may be motivated to share the experience via a geowiki so that others may benefit [4]. A problem with wikis is that there has to be a moderation activity to reject obvious abuse. While these activities may also be performed by noncommercial volunteers, there may also be a vital role for commercial validation services.

This option is perhaps something of a wild card. Could there be a grassroots movement to provide a Whereness "common"? There are already many millions of WiFi hotspots mapped and open source resources to manage the information space. There is also an increasing interest in amateur "mashups"[2] whereby open application programming interfaces (API) from large organizations can be combined to provide some very compelling location-aware applications. It is possible that this grassroots approach might evolve into something of a standard that gathers its own momentum. Open mapping is discussed in Chapter 8, and if this investment in personal time were accounted for conventionally and added to the billions of dollars and euros invested publicly in GNSS, it may be that this approach actually becomes the dominant one. Main industry players may then be relegated to providing the low-value elements and specialist systems integration, while the mass market is effectively free. A caveat is, however, appropriate. The "tragedy of the commons" [5] occurred when overgrazing led to starvation. So although a "free" system may emerge, it may ultimately fail due to abuse and be rescued by a more traditional approach from established players.

If people know their location then why would they not share that information for free or maybe for a reusable token? There is the problem of validation because how would they really know they could trust the positional information? Perhaps the shared location or route is a false one calculated to draw the victim into a false location for some illegal purpose. Really important application users are likely to want to validate the data or at least the identity of the other party providing it.

---

[1] A common is a metaphor for a shared resource owned collectively, inspired by medieval village commons that are open spaces upon which the ordinary people, or commoners, have animal grazing rights.

[2] A mashup is the term used to describe the informal integration of information taken from several sources. Many new user applications in Web 2.0 can be made by people mashing up existing data accessed from specific remote Web sites together with new data provided locally.

3.4 Whereness and the Future Digital Networked Economy

It would be useful to consider how the digital networked economy (DNE) is developing and how it will affect the emerging Whereness business. The term is used by the ICT industry to describe the business of the Internet, which is really about the business of the internetworking of computers. As time advances, more and more devices that were analog, for example, the telephone, Hi-fi, TV, radio, and personal stereo, are becoming digital and effectively becoming computer-controlled appliances that are internetworked. We now consider some of the more important aspects of the future DNE, how it impacts on Whereness, and vice versa.

3.4.1 Falling Cost

The cost of electronics, computing equipment, and software will fall, perhaps more quicky than in the past due to the effects of rapid commoditization and the growth in the Asian manufacturing and software economies. As enterprise software and related issues of security are replaced by more up-to-date versions, the costs, associated with problems of the first-generation solutions, should fall as they get fixed. Whereness will enhance trust and security and should help this trend.

The relatively high cost of personal navigational equipment, wireless enabled computers, and sensing systems will fall and become more affordable. The business case for position-aware business systems thus improves and consumer products will increasingly include positioning-dependent applications.

3.4.2 Web Services

As the numbers of internetworked computer users increase, Web services will become increasingly important. A Web service is an application offered on demand to a user by a service provider over the Internet or an intranet that normally does not need any special software to be installed on the user's computer over and above the normal browser and associated elements such as plug-ins, cookies, and Java support.

All Web 2.0 applications are Web services. The application intelligence is distributed between the browser of the user's platform and the Web server that is providing a service. Chapter 8 describes how it operates and includes details of the principal location-based services such as Google Maps and Google Earth and methodologies such as AJAX[3] and APIs. Of all the developments that are relevant to Whereness, the recent advances in this area cannot be overestimated in importance to Whereness. Recent devices such as the Apple iPhone are allowing

[3] Asynchronous Java Script and XML (AJAX) and the Application Programmable Interface (API) are important aspects of Web 2.0 and will be explained in Chapter 8.

Web 2.0 Web services to become truly portable and ubiquitous, and it is very likely that these devices and others that follow them will become the main way by which people experience Whereness when they are out and about.

More people will be able to participate in the emerging market for positioning-aware systems since the cost and complexity of locally installed applications are avoided. There will be greater business associated with all the elements of Whereness that becomes largely invisible to the user and becomes part of many Web services rather than a specific end in itself. For example, a system to manage business meetings and venues would be able to take account of the positional aspects travel, logistics, and carbon emissions.

3.4.3 Geographically Tagged Information

While basic networking, infrastructure, and software businesses may become less profitable, there will be growth in the information networking business. Whereness may lead to some very profitable "information spaces"[4] concerning geographical information and related information. For example, business information could be "tagged" on a map, which would be delivered by a Web service, to show the location of facilities to customers. Web portals and homepages for some organizations may center on maps with geotags. Some of these tags may have a real-time aspect. For example, dynamic tags may show the location of deliveries or actual progress of public transport.

3.4.4 Information Commons

Open systems and information commons will increase and erode the business of traditional information providers. Positional related commons will emerge, for example, OpenStreetMap [6] for map creation, Wikitoid [7] for registering topographic identifiers (TOID), Wardriving [8] for mapping WiFi access points, and Place Lab [9], that takes the mapped Hotspots and derives location from any signals received locally.

3.4.5 Trust Support Services

Established computing, networking, communications, and information organizations will suffer increasing competition and will move into new value-added areas providing trust, quality control, payment systems, and digital rights management. Smaller and newer organizations and open commons will use these services to increase the trust of customers.

---

[4] An information space is a computer science and mathematical term to describe all the information relating to a single subject.

3.4.6 Opportunities as the Scope of the Internet Increases

Research by computer scientists into ubiquitous computing will lead to new products and services as the "Internet of Things" gathers momentum. There will be several orders of magnitude increase in the numbers of networked digital processors that will generally be fitted with sensing capability and wireless communications. New interfaces based on objects or artifacts will be common.

Whereness will benefit from a very rich sensory environment that can be leveraged without significant extra investments (e.g., intelligent building environmental control systems).

3.4.7 Simplicity

There will be a demand for simplicity from users who will pay extra for easy-to-use applications that need no learning or training [10]. While the internetworked world will increase in its complexity, this will be hidden from the users through value-adding service providers who will increasingly automate interfaces and applications.

Many problems that involve guidance, logistics, transactions, monitoring, and control can be automated using embedded ubiquitous positioning, to the point that the technology appears to disappear. It will be taken for granted that everything will have knowledge of the position and status of every relevant other thing. Automatic rules will be embedded to initiate action should any foreseen circumstance arise.

3.4.8 Wirelessness

Wirelessness [11] will increase as new standards such as WiMax are used to bring broadband access to buildings, and within buildings WiFi will be supplemented by newer standards such as ZigBee, to communicate with sensor networks and other building plant. The scope for mobile and hotspot operators to use wireless networks in a positioning and sensing mode will increase and lead to new value-added services involving Whereness. These have potential to be more profitable than the primary access communication for which the infrastructure was first deployed. As environmental concerns deepen, there will be a rich sensory and control environment within most buildings. These systems will be useful to aid general positioning but may also require positioning information to help manage energy resources effectively.

3.4.9 Whereness as a New Utility?

Telcos, ICT organizations, and infrastructure providers will increasingly move into converging systems and into the provision of value-added services.

Ubiquitous positioning may become more profitable than some of their current services. Whereness may become viewed as a new utility in its own right.

3.5 Businesses Relevant to Whereness

There are many general areas and trends in the future digital networked economy that could be important. So what specific businesses and systems could contribute to the convergence and what are their trends? The following sections cover the most important.

3.5.1 Global Navigational Satellite System (GNSS) Receivers

The increased use and lower costs of Global Navigational Satellite System receivers generally and their integration into portable computing platforms will increase the business of the chip-set and original equipment manufacturers (OEM) and the appeal of the end products. It is likely, however, that the software-defined radio[5] will perhaps place some of the complex processing out of dedicated GNSS chips and place it within the software of the end device (along with any other wireless processing). Standard high-end general purpose chip makers will increasingly be placing software radio subsystems within their standard products so that most general purpose processors will contain a wireless, or at very least, some of it.

As more systems become available, multiband GNSSs will become more common, adding to the opportunities for the receiver providers. Within about a decade, we could expect to have at least three systems available (enhanced GPS, Galileo, and revamped GLONASS). Actual availability will depend on the health of relevant economies and politics rather than on technologies.

A wildcard concerns the future of spacecraft. Until recently it was not difficult to predict the enormous cost of placing platforms in orbit but the advent of space tourism just might change the economics of space. This is unlikely to be evident for at least a decade until the second generation of tourist craft is launched, assuming the first generation is a success.

Another wildcard concerns the future of high-altitude platforms (HAP), which might be an alternative to GNSS. Very high altitude solar powered airships or aircraft are being researched at present. They may, if successful, be used for communications, remote sensing and imaging, energy generation, and navigation.

3.5.2 Cellular Mobile Radio Operators

Basic voice and data services are now highly competitive. Future revenues will be increasingly dependent of value-added services, especially those involving social

[5] Software-defined radio is described in Chapter 9.

groups. These services will be supported by new handsets that are becoming more sophisticated, attractive, and capable. New capabilities will include hotspot wireless, GNSS receivers, sensor technology, and new display technology. A novel possibility, based on recent research, is the use of camera phones to provide position and location (see Section 7.4.2).

3.5.3 Wireless Hotspot Operators

Increased numbers of public, private, and hybrid WiFi hotspot access points will lead to new positioning services. Hotspots can be used to expose the location of user sessions in two ways: first, via the network operator from knowledge of who is logged on to each base station, and second, by users harvesting the broadcast ID information and comparing it with maps. The amateur wardriving activity [8] of logging hotspot location with GPS coordinates (which is illegal in some places) may become a mainstream business activity.

3.5.4 Groupware

Corporate e-mail and groupware already include the functions of meeting organization, and portable groupware devices such as Research in Motion's Blackberry, and are becoming more popular. Groupware will be improved once Whereness is offered.

Consumer mobile devices will have similar services but are more likely to be linked to social networking Web service portals. In the longer term there may be a move for corporations to use modified versions of the social portals, since employees will be highly trained in their operations. Advanced billing, security, and session control will enable the same (domestic) device to be used for corporate use, thus avoiding the need for people to carry two mobile devices. The Whereness services can thus be offered to a person according to two personal profiles, one corporate and the other domestic.

3.6 Intelligence

Assuming customers have access to time and place information, it is likely they will want more, especially if useful advice about actions and activities can be provided automatically by some sort of machine intelligence in a simple and clear way. Value-added services of this nature will probably be the areas that will be profitable in the future digital networked economy.

3.6.1 The Invisible GIS

The temporal aspect can be immediate, involve past actions, or concern predictions but generally concerns events. A positioning system may be associated with a map to show where the subject is, was, or will be, but then generally be associated with an action to do something. If systems are automated then the "map" may be present as a geographical information system (GIS) but perhaps not visible to any person. Some examples include:

- My colleague may be in a meeting room on a different floor of the building so a text message is sent to my mobile phone to tell me the details;
- My child did get to school on time so no action is needed (this time!);
- The nearest ambulance would be nowhere near the accident in 15 minutes because there is a major traffic jam ahead, so another vehicle must be dispatched urgently from a different depot.

There is a three stage process involved, which in human terms is sometimes expressed as shown in Figure 3.1.

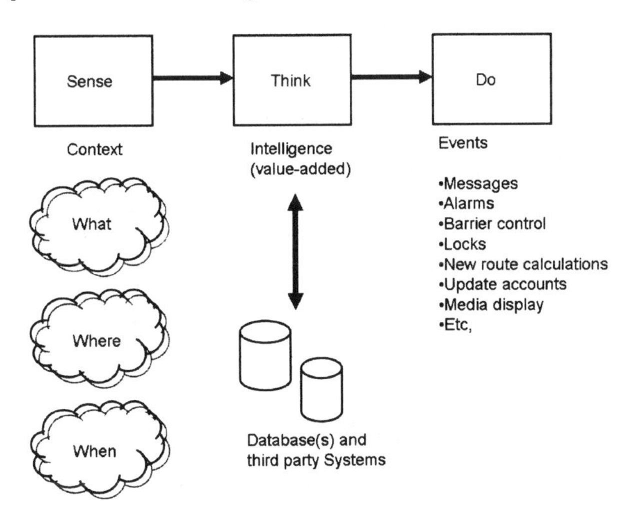

Figure 3.1 The three stages of automation are sense, think, and do.

Automation in the "think" stage can be achieved using artificial intelligence (AI) software and associated databases to infer situations and then the "do" stage initiated either directly or indirectly via human intervention. All stages are very likely to be dependent on communications and computation that are becoming distributed. Some tasks may be performed locally or may be offered via a remote server or may use a hybrid approach. Businesses can generally save on infrastructure costs if the intelligence can be shifted to the edge of the network into the user's device.[6]

From a business perspective this intelligence is very important. In addition to converging positioning methods and offering security and identity management, it could be the main thing that customers will be paying for. The more difficult a problem is that can be solved automatically, the more people would be likely to pay for the information concerning a solution. Some of the more important common intelligence requirements are now discussed in more detail, including:

- Routing information such as messages, multimedia, and machine commands;
- Accounting units of value such as money, energy, and personal exercise;
- Route guidance for people and vehicles;
- Controlling systems including access, equipment, assets, and radio systems.

These are called common capabilities because they support many different applications but have a very similar underlying function. The databases, computing platforms, and software for each of the capabilities can be very similar and thus reusable after appropriate configuration. They differ, because each type needs to interact with other associated information spaces. Some applications may use several or all of these capabilities, and it is likely that some operators of services may subdivide the capabilities. Some details of each common capability class are now given with some typical application examples and details of associated information spaces.

3.6.2 Routing Information

One of the main applications in the original ubiquitous computing vision is the idea of media "following" the user's movements. Assuming there is a location system in operation (for now it does not matter how this is being done), as the user needs to interact with some media, it will appear on the screen or other display close by. Essentially there is a decoupling between the terminal and the application and a distributed computing model is assumed.

Another area of interest would concern location-based services for mobile phone users who can have messages routed to them based on their location. A set of rules would be followed according to the location of the user. Messages would be sent, stored, returned, or deleted according to the rules. For example, a rule could be set "return all messages from my work colleagues when I am known to be at my holiday home." Associated information spaces are user names, addresses, diaries, calendars, e-mail, SMS, and other messaging systems and all manner of information resources.

---

[6] Web 2.0 and its use of AJAX, as described in Section 8.3.6, is an example.

3.6.3 Accounting Units of Value

Units of value would normally be money, either credits or debits that are added or subtracted when the user moves into a zone, along a route, or within an area (often bounded by what are know as geo-fences). Tolls, entry or exit charges, fare collection, and movement charges can thus be collected automatically [12]. There are, however, other units of value that could also be accounted in a similar way to money, since money is only a form of information in the electronic economy. An emerging area could be the collection of energy use or pollution emissions to help manage climate change. Another emerging area could be about collecting units of personal exercise to help manage obesity, accounting for the calorific value of steps taken and at what speed and gradient. More futuristic could be the accounting of virtual gaming credits and scoring for future sporting activities.

The informational processing looks for charging or accounting events and then follows a set of rules to increment or decrement the user account. Associated information spaces include user accounts, pricing tables, billing systems, and scoring systems.

3.6.4 Route Guidance

Time and movement information are combined to provide tracking and movement prediction based on a dynamic movement model for the subject of the service. This application is about routing people or objects rather than about routing information (although the latter can help inform the former). Positional information need not be continuous but can be inferred provided enough values are known from spot measurements of position. Routes can be found from maps and then can be automatically followed. Messages can be delivered to help both drivers and pedestrians who are following routes and who may need to reroute based on the contents of the message. Alerts can be generated if deviations are detected and updates provided if circumstances change. Since routes that are traveled are only available in certain places, any positions detected that are not on the route may be indications of new routes.[7] The route guidance system can also be used to build maps as well as follow them. Time estimates can be calculated and optimization services included. Geo-fences can be monitored to provide event information when they are crossed.

Associated information spaces include GIS containing map overlays with attributes such as post-code data, postal addresses, traffic regulations, traffic conditions (current, historic, and predicted), legal information relevant to customs, hazardous loads, areas where offenders may or may not be, and many other geographical information sources.

---

[7] This is known as simultaneous location and mapping (SLAM) and is discussed in Chapter 9.

3.6.5 Controlling Systems

This is an emerging area but mainly concerns machines communicating with machines, one or more of which are capable of being mobile. Examples include barriers that are operated automatically and triggered by people or objects moving into certain areas, tracking the progress of goods in warehouses and supply chains, baggage in airports and transport systems, and maintaining the security of assets by checking that they are where they should be located.

Mobile robots are an important future area for automatic control since their position is a key element in both safety and efficiency. Wireless network spectral management, although basic at present, will increasingly be able to benefit from more efficient control based on accurate positioning (see Section 9.1.3).

3.7 Autonomy or Centralization?

In some positioning systems such as GPS, the determination of location is based on local or autonomous processing. Although the common infrastructure provides the base information via a one-way broadcast from the satellites to the users, it has no part in the sensing or processing operations that are entirely local to the user. Virtually limitless numbers of GPS receivers can therefore be operating without any reduction in the performance of the system. In other words, these systems scale well.

In contrast, many other positioning systems work in the opposite sense where the processing is done by the infrastructure and the user role is minimized. For example, an RFID system in a warehouse that is fitted with a network of tag readers may be used to track a tagged consignment of goods on a pallet. In this instance, the local tag plays no part in the processing other than communicating basic ID (identification), but it is the network that is doing the information processing. There is an upper limit to the number of tag events that can be handled by each tag reader (i.e., the system does not scale infinitely and using more tags may imply more tag network investment in order to read them effectively).

There are clearly some trade-offs and these are important to consider when building a business case for any positioning system, ubiquitous or otherwise. Overall, GPS seems a better approach until the prices of a tag and a GPS receiver are compared. A tag is 2 or 3 orders of magnitude cheaper and also may not need any battery and can be extremely robust. Overall there is much to be said for both approaches so the ideal may involve a hybrid approach. There is a general trend to push intelligence to the edge of networks but still maintain certain functions centrally. In general, the autonomous approach seems more useful in terms of scale but at the expense of more complex, expensive, and power-hungry mobile units. On the other hand, centralization puts more power in the hands of operators and may be more attractive in niche markets and markets for early adopters.

3.8 Quality of Service

There are a number of important quality issues. First and most obviously is accuracy of positioning, which can vary greatly. Second is the availability (i.e., where positioning is possible or not). The main issue for Whereness becoming successful and truly ubiquitous will be to converge enough technology and services to fill all significant coverage gaps. Third, the time it takes[8] to provide the temporal-spatial information is important. If services are to be reliable and profitable, then sensing, computing, and communications must be fast enough to be useful. Instructions that are too late are a common failing in mobile guidance systems.

3.9 Privacy, Trust, Security, and User Profile

It seems unlikely that the full potential of Whereness will be realized unless there is consent and trust between the system and the users. To ensure that consent and trust are forthcoming, the businesses providing the equipment, services, and support must be trusted, so it can be expected that trusted brands will be at the forefront of Whereness.

One of the biggest problem areas with anything to do with location and tracking is the question of privacy. "Would I want my boss, partner, parent, insurance company, police, or government to know where I am or where I was at some time?" Very often the answer to that question would be no, since it is none of their business (in my opinion) and because it is limiting my freedom in some way. On the other hand, it would be reasonable perhaps, or even desirable for health and safety reasons, to be monitored while at work, particularly if in potentially dangerous situations. With close family and friends it would depend on social circumstances and with the other agencies it would depend on legal issues (i.e., a matter for courts, contracts, and lawmakers). In all circumstances, it is an issue of trust that needs great flexibility and ideally should be under the control of the user via a personal profile.

3.10 Whereness service provision

Having now considered the important business issues, is there a methodology to piece together this jigsaw into a coherent whole that could be offered as a general Whereness service? Figure 3.2 shows one way in which it might be achieved and used as a starting point for a potential Whereness operator to consider.

---

[8] The term "latency" is commonly used to describe the delay inherent within systems.

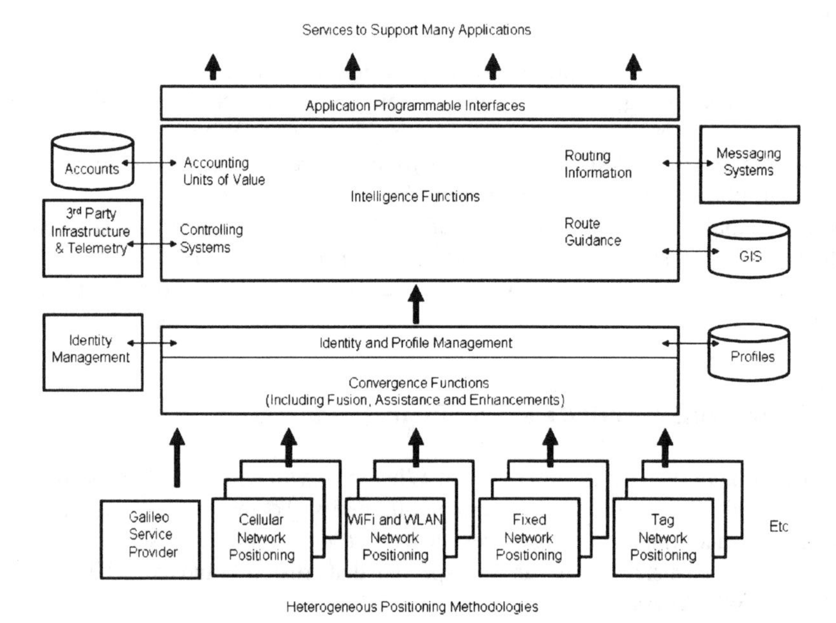

Figure 3.2 Methodology for the provision of general Whereness services.

At the bottom of the diagram are a series of positioning systems, all, some, or none of which may be owned by the operator but all of which have an agreement to pass positioning information. Convergence functions for each of the systems convert information into a common format. Above convergence comes the trust layer managing identity and service profiles for all customers. External systems may be used to manage identity operated by other parties. A common interface is then presented to a set of intelligence functions that perform the value-added functions and that rely on many external systems and data sources, in particular a GIS. Applications are then supported via a common API that may become standardized across the emerging Whereness industry.

This approach is flexible and to an extent future proof. Extra positioning systems can be added and new convergence handlers written without changing the internal common approach. The API can be extended or even better, written so that the future functions are already present but not presented until operational.

However, it should be remembered that this diagram makes no statement about the colocation of functions, as it is a logical diagram, and it is very likely that a very distributed approach will be taken in the future with many of the functions actually running on user hosts and not central servers. Centralized functions are there for support, convergence, and occasional synchronization rather than for continuous sessions.

In the following chapters, the technologies of positioning and related systems will be discussed but it might be worthwhile to refer to this diagram from time to time to see the bigger picture. In the final chapter, this topic will be revisited but from a different perspective: that of the information spaces, using the principles of the Semantic Web to bring together the fragments of information needed to create intelligent services and encoding them according to an ontology[9] for ubiquitous positioning. Given the complexity of service provision illustrated in Figure 3.2, it is envisaged that Whereness service provides may be able to save expensive software integration by automating the process by adopting a Semantic Web approach.

3.11 User groups

There are many ways to group services according to users, customers, business sectors, or organizations. Figure 3.3 shows one useful method that separates personal life from work life and that shows some of the most relevant business sectors to services. Starting with a person, that person is likely to have social and work-related requirements. Although a common system might be used, the services are likely to be needed separately (in the same way that a person might have two mobile phones, one for work and the other for home).

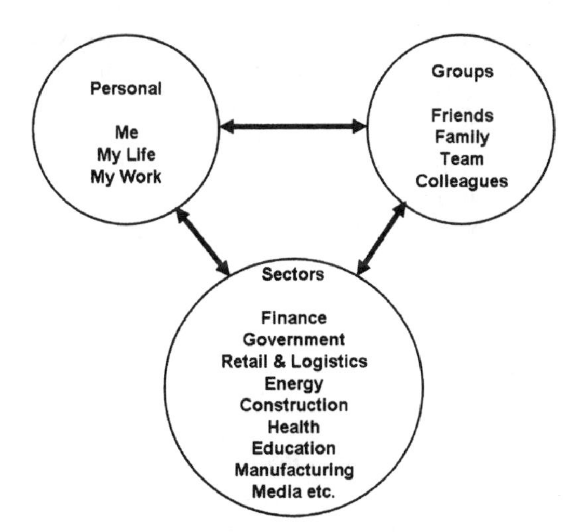

Figure 3.3 Classes of users.

[9] Ontology is a key concept of the Semantic Web and is explained in Section 9.3.1.

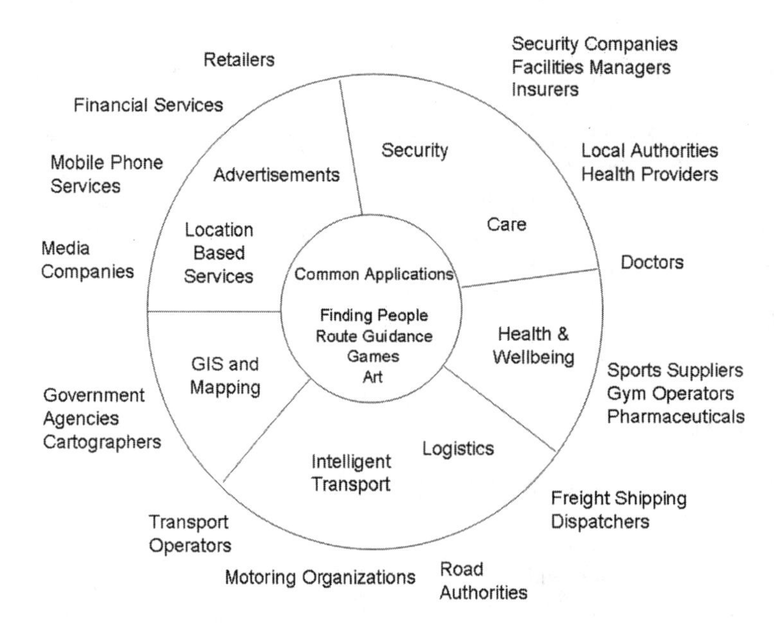

Figure 3.4 Whereness user groups.

Some businesses will benefit from Whereness more than others but in general there will be some applications that are common to all. Similarly, some consumer groupings are more likely than others to use the social side of Whereness while others may benefit more from the care applications. Each of the more important applications of Whereness will be discussed in the following chapters. Figure 3.4 shows these applications clustered according to business sectors. In the center of the diagram the personal applications are clustered that are not sector specific and would be provided directly by a service provider to individuals either personally or via their employer.

3.12 Summary

Whereness has an important part to play in the digital networked economy both now and increasingly in the future. ICT costs are falling so the relatively high costs of some elements of Whereness are becoming less of a problem. More computer applications are being offered as Web services as Web 2.0 makes steady progress, and map-based interfaces are some of the most popular, which is allowing users to geotag information.

Information commons are a new and potentially disruptive business model and the open aspects of Web 2.0 are leading to a very different approach to mapping applications that are often used to display positioning information. As the scope of the Internet widens and more devices become networked by wireless, the resulting infrastructures help Whereness. Several business areas of relevance were reviewed including GNSS, cellular radio, wireless hotspots, and groupware.

Simplicity and automation are also important to provide new position-aware value-added services that may be demanded by customers and be very profitable. The importance of software intelligence was highlighted and the automatic systems operating in the background of Whereness services were reviewed including the invisible GIS (i.e., mapping applications that are not displayed), accounting value by position, route guidance, and system control.

Several other important topics were discussed, including the importance of trust, user profile, and security. Finally, an overall generic logical architecture for Whereness was suggested and diagrams provided to help segment the services it would provide.

References

[1] ITU, *World Telecommunications / ICT Indicators Database (10th Edition)*, 2006.

[2] Quiktrak, *Land Based Radiolocation*, http://www.quiktrak.co.uk/.

[3] Wikipedia, *Free Content Encyclopedia Project*, http://en.wikipedia.org/wiki/Wikipedia.

[4] Erle, S., Gibson, R., Walsh, J., *Mapping Hacks*, Chapter 1, "Mapping Your Life," pp. 1-53.

[5] Garrett Hardin, "The Tragedy of the Commons," *Science*, Vol. 162, No. 3859 (Dec 13, 2009), pp. 1243-1248.

[6] *OpenStreetMap*, http://openstreetmap.org/, Jan. 2008.

[7] *Newcastle University, Civil Engineering and Geosciences*, http://www.ceg.ncl.ac.uk/research/geomatics/projects/wikitoid.htm, Jan. 2008.

[8] *WarDriving.com*, http://www.wardriving.com/, Jan. 2008

[9] *Place Lab*, http://www.placelab.org/, Jan. 2008.

[10] *Simplicity at the MIT Media Lab*, http://simplicity.media.mit.edu/, Jan. 2008.

[11] Dennis, R., Wisely, D., "Mobility and Convergence," *BT Technology Journal*, Volume 25, No. 2, Apr. 2007.

[12] *Transport for London*, Oyster online, https://oyster.tfl.gov.uk/oyster/entry.do, Jan. 2008.

Chapter 4

Current Whereness Applications

This chapter looks at what is possible to do today and in the medium term. Each application is described and the positioning element highlighted. The benefits and weaknesses in the application are discussed to help potential operators and users get a balanced view of what may be possible. Most technology can be abused and radio technology is particularly vulnerable to external interference and can suffer from poor coverage. The basic requirements are discussed and links to other systems are highlighted. Chapter 5 is more futuristic and concerns things still emerging from the research phase whereas in this chapter the barriers are more about the commercial viability of applications and systems.

4.1 Mobile Information

The first application to use position-dependent information delivery was traffic information services for drivers. The are many ways to deliver position-dependent information, for example, a simple approach is where a driver calls a number on a mobile phone using a hands-free facility and the approximate location is passed from the network to the traffic service provider to select local traffic reports dependent on the identity (and hence approximate location) of the local cell base station that is handling the call.[1]

Another simple approach that has been used is to broadcast traffic messages as data codes using broadcast FM local radio's RDS system (usually used for automatic tuning, station identification, and simple spoken road information). A message set has been standardized as the traffic message channel (TMC) that encodes standard messages into a 16-bit code. There are 2^{16} possible preset messages and a second 16-bit code to represent location (i.e., there are 2^{16} possible preset locations in the region). A vehicle location system can then select the location codes that are relevant and display only the local relevant messages.

[1] This approach is know as Cell ID and it is explained together with more complex cellular position in Chapter 6.

Although somewhat inflexible, the RDS-TMC system [1] is very cost-effective since it can be integrated into the vehicle radio. The advent of digital radio (e.g., DAB) is a potentially more flexible approach.

The benefits are simplicity and ubiquity but there is no obvious way to create a revenue stream other than treating the messages as an attractor to the station to increase its audience numbers and advertising revenue, if commercial.

The section covering mobile advertising is probably a more likely way for simple location-dependent information services to progress.

4.2 Dynamic Route Guidance

Route guidance equipment (DRG) is now standard equipment in vehicles based on GNSS and digital map matching, but it is only recently that the routes suggested by the route-finding software are becoming able to take into account real-time information such as unexpected traffic delays.[2] This dynamic approach to route guidance has been a long time coming and is still primitive in the way it operates.

In the late 1980s and early 1990s there were research programs and projects[3] that showed that DRG could be used for much more than personal guidance, and could be part of an overall system that could lead to a reduction in overall traffic congestion and also have potential to manage the environmental impacts. Today, cellular radio can be used to get simple traffic updates collected by a variety of methods (e.g., roadside traffic sensors based on cameras or inductive loops buried beneath the road). The original vision of the experiments is still a valid prospect and concerns the idea that the vehicles act as traffic probes (collecting what is sometimes known as floating car data or probe data).

Rather than merely feed back the raw information to other users and thereby create "rat runs" into inappropriate residential neighborhoods, the experiments were also about a centralized journey planning system where local traffic management authorities would dynamically route some traffic one way and others a different way to balance the road network congestion (using the same body of theory that communications and Internet data traffic engineers employ). In the early 1990s the main objective was to save journey times but today and increasingly in the future, DRG is more likely to be part of a scheme to manage carbon emissions and perhaps be linked with dynamic road pricing. Urban traffic management at present is a macroscopic management activity. Accurate vehicle positioning with integrated mapping, based on an enhanced satellite navigation unit (or sat-nav) with near real-time communications, could be part of a microscopic system where the progress and effect of every vehicle are controlled at all times, or at least in important urban areas.

---

[2] Chapter 8 has details and references of some service providers using Web 2.0 for mapping and route sat-nav manufacturers such as TomTom who add dynamic services to basic stand-alone products.

[3] IVHS America (Intelligent Vehicle Highway Systems) in the United States and DRIVE, part of the EU's Second Framework Research Programme.

Two of the early experiments illustrate the two different approaches to centrally managed dynamic route guidance. The Ali-Scout systems from Siemens [2] used a network of local infrared traffic communications beacons that acted rather like electronic sign posts. The other system was the System of Cellular Radio for Traffic efficiency and Safety (SOCRATES) [3] and was a collaborative project of the E.U. Framework research programs. It was based on cellular radio and autonomous navigation. Both systems were before GPS was generally available but their performance was in some ways better as GPS coverage was never an issue. Other projects, such as PROMISE [4, 5], pioneered the portable personal information terminal for travelers who were not driving.

Ali-Scout and SOCRATES used a combination of magnetic compasses, vehicle odometers, and the all-important digital vector maps, which treated a route as a set of vectors connecting a set of nodes.[4] It is useful to look at these systems in more detail, as the way they worked is now much closer to the way indoors person-based systems would work and thus sets a framework of a universal ubiquitous positioning approach to guidance.

The Ali-Scout system used modified traffic signals, which in addition to the normal colored lights had extra dark lamps that emitted infrared data. At each node, usually a city junction, the data message broadcast to any passing vehicle was the vector map fragment to take the vehicle to the next node in the journey. The vehicle communicated its destination to the beacon, which responded with the appropriate map fragment personalized according to the journey destination. The central traffic computer did not need to know every vehicle's journey, but worked rather on a standard set of possible destinations. It was easy to collect vehicle delays from the vehicles since they had followed a known route from the last beacon and because the journey segments were timed. The central computer thus continuously calculated the most efficient route and node scheme for an entire city based on real-time measurements and any other a priori information.

The SOCRATES system did not need the map to be communicated since it was carried in a digital database on a CD-ROM in the vehicle. Cellular radio with an embedded data system was used to communicate road link impedances to the vehicles from the central system. In turn, the central computer collected a global view of congestion from the vehicles using the data link since each vehicle measured link impedances locally. Routes could thus be calculated in the vehicle autonomously, taking into account real-time traffic conditions.

The map included historic link impedances, but these were supplemented by the real-time centrally calculated information that was regularly updated. By adjusting the impedances, the central computer could influence the routes taken. This system was more autonomous than the other, had more expensive in-vehicle equipment, but did not need a beacon network that required expensive landline-based data circuits.

---

[4] Chapter 8 explains the significance of digital vector mapping and a brief example is shown in the Epilogue.

The trouble with both experiments was that the costs were too high for commercialization and it is only now (given the advances in ICT) that such systems could again be contemplated. A key feature of these systems was the hybrid nature of the intelligence, with local guidance being vehicle borne but the strategic guidance being centralized (and under control of the authorities). A number of linked databases were used together, and it is worthwhile to explain their roles since the approach is still valid for the future in other types of guidance services (for example, for nonvehicle-based individual travelers or guiding people around facilities).

4.2.1 Static Data

The static data mostly concerns the vector map that does not change very often. It might also include attributes such as speed limits and road charging schedules (per unit distance, per time of day, and per class of vehicle). It is the static data that the positioning and guidance system uses together with fused sensor data. The data can either be carried within a moving device and cover a large region or be served on demand by a wireless communications session.

4.2.2 Historic Data

Historic data includes traffic flows averaged over a long period and can take the form of a set of road link impedances per time of day, including weekly, monthly, and any other recognized patterns. The information may already exist from traffic census activities or be derived from the data collected from vehicles (averaged and made anonymous).

4.2.3 Semidynamic Data

Some sporadic information about traffic delays might be known well in advance, such as planned roadwork or a major sporting event. These impedances can be added as a nonreal-time overlay to the historic data.

4.2.4 Dynamic Data

Dynamic data is real-time data collected from the vehicles or road sensors and reflects the link impedances in real time. Its quality will vary, as the sample size will tend to vary according to the number of vehicles feeding data back to the central management system.

4.2.5 Predictive Data

If a lengthy journey is planned the link impedances used to calculate the route must be predicted, especially those that are going to be traversed in several hour's time. Therefore, when a route is requested it is predictive data that is used to calculate the route. Clearly, there is scope to add value to the historic data and to charge for this enhancement.

4.2.6 Combining the Data

A major challenge would be to find ways to combine all these data sets, particularly if they are all controlled by different agencies in different places. A computer model is needed to continuously produce link impedances for several time horizons. It may also be possible in the coming years to make more progress using the methodology of the Semantic Web to collect fragments of relevant information from a variety of sources.

4.3 Auto Payment Systems

Road user charging (sometimes called road tolling, road pricing, or congestion charging) and pay-as-you-drive insurance are both examples of automatic applications that involve the collection of charges. Accounting can be by distance driven, time of day, type or class of road, and can include details of actual road links traversed. A national or international system could provide the platform for most other ITS applications and a good number of more general LBS applications as well. There are many niche approaches at present, including using microwave tags and video surveillance. The latter is often needed in addition to help with enforcement. To overcome the split between public and private applications, a public-private partnership may be possible (which is common in other infrastructure projects in some administrations).

Markets will thus span many sectors, including local and city authorities, national governments, insurance companies, and other ITS providers, and may include additional value-added services, such as dynamic route guidance.

The benefits are that it is the most flexible of all methods of transportation taxation and also for insurance premium collection. Automation will minimize manual overheads. It could be linked to dynamic speed limits and their enforcement, but this could be a serious barrier to public acceptance, which is likely to be problem, in common with any new tax. Green awareness might, however, mitigate since the extreme flexibility of pricing is likely to be the most effective and fair way to reduce unnecessary emissions.

Other benefits are a potential modest increase in road throughput efficiency if linked to a route guidance regime. It could be used to counter crime and terrorism but there could be resistance to this also.

Weaknesses include a relatively high cost of in-vehicle equipment, radio coverage, and interference issues. Failure problems may be an issue for a national government system so a multimodal positioning technology approach may be needed.

High accuracy is needed, <10m, with second-by-second positioning measurements. Sporadic hotspot wireless access may be needed (e.g., at refueling stations) for updates of local charging schedules and downloads of map segments driven or credit units. There is a requirement for legal enforcement and assurance of system performance and contingency plans for any system failures. An alternative partial approach may be offered by use of extensive video surveillance and (number plate) recognition systems. Ideally a dedicated system is needed with quality guarantees, which is one major advantage of Galileo. There is a need to include GIS for toll and premium geo-fencing. Although there is no specific requirement for real-time communications, it may be included as part of a wider ITS service set.

4.4 Parking and Ticketing

Automatic parking charges for motorists and fare payments by public transport users will be possible provided both vehicles and people carry the appropriate Whereness technology (see Section 3.6.3). There may, however, be a need for manual fallback systems. Automation implies the ability to "just travel" without any thought of the payment systems, which should be, ideally, entirely automatic. Systems could be extended to other forms of ticketing such as entertainment and artistic performances, for the granting of access to buildings, and payment of access charges for events and to facilities. Markets include:

- Operators of vehicle parks;
- Operators of public transportation (bus, train, tram, taxi);
- Owners of buildings with charges;
- Facilities managers with charges;
- Box offices for theaters, cinemas, concert halls.

Benefits are lower costs by using a common platform across industries and thus using the economies of scale across many businesses. Local authorities will be encouraging carbon emission's reduction by promoting public transport with "through ticketing" and easy modal shift. Facilitation of park and ride schemes would be included. There will be less need for users to carry cash and a plethora of cards and tickets, thus reducing the scope for fare cheating and crime. Systems could be extended (with suitable validation safeguards) into car sharing schemes by matching the position of potential passengers with partially empty vehicles.

Weaknesses will include gaining the agreement from multiple authorities. Users will have to be persuaded to carry "technology" that could be as simple as a universal RFID tag or as complex as a smart phone with GPS. There may be public resistance due to Big Brother concerns and impatience if the systems are not engineered robustly (a potential problem, given the dependence on wireless).

Operators stand to lose revenue if the system is prone to failure, cheating, and poor enforcement.

Very accurate and exacting point positioning in real time is needed (<1m at barriers but otherwise <10m is adequate). Low latency is required. Multiple technologies can be converged, including existing RFID technology, smart card (and virtual smart card) solutions, cellular positioning, and GNSS. There is a need for a GIS and integrated timetable and fare schedules. Intelligence software is needed to predict problems for individual travelers in real time, to alert them using various messaging methods, and to offer solutions and guidance services [6].

4.5 Emergency Calling

Before the advent of mobile networks, the plain old telephone service (POTS) offered a very basic location service to emergency control centers. Although a driver may have had to walk for some time to find a fixed telephone, at least when the emergency call was made to the network operator's emergency call center, the operator could see where the call was coming from geographically. It is worth noting that (some) fixed communications systems can offer a very high level of positional accuracy since the terminal positions are fixed and known. Unfortunately, with the advent of Voice over IP (VoIP) telephony, the association of line with terminal has been abandoned. VoIP phones (or PC applications) now require an external Whereness service to restore the positioning functionality.

Digital mobile phone networks provide various levels of positional information that can be associated with voice calls, messaging, or data services. The most basic uses the identity of the base station handling the session. This Cell ID can then be referenced to a known coverage area (perhaps a set of postal codes or some other geo-fence). In Chapter 6, more complex arrangements will be discussed but here is a description of Cell ID.

The operation of the underlying network mobility management system for the dynamic allocation of radio channel resources requires that a database system track each terminal's approximate location based on the local fixed station with the best quality of service. As a terminal moves, the logical channel (handling the call) is reassigned automatically during a handover process from cell to cell along the approximate geographical track of the user. In some third-generation systems this approach is somewhat more complex, whereby several base stations can be handling the call concurrently and be involved with a "soft" handover process. In general, however, it can be said that a particular geographic area represents the coverage area or "cell" that has a unique identity, or ID. The size of the cell will vary according to environment from hundreds of meters in dense urban areas to tens of kilometers in very rural situations. Satellite cellular services may be even more inaccurate, as spot beams illuminate hundreds of square kilometers. The great advantage of Cell ID is that it is information that is virtually free to the operator. Network signaling systems sometimes contain the information as a given information field and only a small amount of development effort is needed to make the information more widely available. Regardless of any other services available, an emergency call service with Cell ID is normally implemented.

It is easy to associate an emergency voice call with its Cell ID and to display it on a map using a GIS. Things become more complicated when calls are received from different mobile networks and even more difficult when automatic text messages arrive (somewhat later), triggered automatically by air-bag inflations. Telephone calls and text messages have a calling line identity (CLI) that can help to associate calls with callers, but this can be withheld. Much work is needed to converge all the various information sets and networks that carry them. As new types of networks and technologies are rolled out, the need for convergence increases and the value of the convergence also increases.

Emergency calling is very important since it involves safety of life and is thus important to public licensing authorities and to most end users since it involves personal and workforce health and safety. In all ITS and LBS scenarios, the case for emergency calling is the most important. As extreme global weather conditions and security concerns increase, emergency alerts with positioning will increase in importance.

Emergency calling can also be two-way. Outbound calling plans can be made to mobile users, if positions are known, to alert them of impending disasters. In some places the old fixed public sirens have been replaced by a GIS-controlled reverse emergency call system.

4.6 Tracking and Logistics

This application assumes that some important objects or artifacts that are largely static or stationary can be moved. Monitoring is important because of their value, operational role, or composition. An object's location is monitored so that alarms can be raised if it is moved into the wrong location at the wrong time. For example, safety equipment can be checked to be present within a defined area. A group of things can also be monitored so that if they are associated together in the wrong way, alarms can be raised. For example, if only a certain number of containers can safely be stored together, an alarm can be raised if extra containers are added. If a number of things are needed together in one place, the service can check or highlight if the wrong item is present.

Markets include industries and operations with expensive or important plants where security and safety are important. High-value asset monitoring in general office environments would be a very wide market and could include portable computers, in-strictest-confidence documents, money, and legal documents. Life-saving hospital equipment and controlled substances could be monitored and hazardous material (chemical, biological, and nuclear) in government-related agencies. Banks and the supply chain of precious metals, gems, and bullion could be tracked.

Benefits would be lower costs for automated record-keeping for health and safety conformance, a reduction in accidents and crime, improved utilization of assets, and a lower wage bill for human security resources.

Weaknesses could include an overreliance on wireless technology that may be subjected to denial of service attacks (e.g., jamming), accidental radio interference, problems due to radio coverage and occlusion, and battery life for powered security tags.

Performance requirements are exacting due to the criticality of many of the objects being monitored. Very high positional accuracy (potentially down to <1m in three dimensions) may be needed in some circumstances. Low latency is required if health and safety issues are at stake. Some systems do not need, however, to be in constant communication and a position audit (perhaps) every hour may be adequate.

At present, positioning and location-based services are a niche market but the future could include scenarios where ubiquitous positioning is the key service to keep a free society functioning in the face of increasing terrorism.

There is some overlap with established security systems such as networked intruder alarms and video surveillance systems. The addition of extra intelligence to a basic surveillance system may be needed, however, to perform a positional check at predetermined times and to react (according to a rule set) according to what has been detected. Systems to be converged include RFID tag systems, video surveillance, peer-to-peer communications with positioning, some long-range telemetry tags based on GPS, WiFi, UWB, and wide area radio (e.g., GSM, SMS, and GPRS).

4.7 Mobile Advertising

Mobile advertising is already a well-established part of mobile transacting, buying, selling, and advertising. If, however, the advertisements are triggered by location of a user or group of users, then more selling opportunities of a real-time nature can arise. Proactive calling and push-messaging to mobile devices can be used to alert potential customers to offers that are nearby, and maps and guidance services may be offered to help customers to find the retailing facilities. Although some aspects of these services have been used in Japan for some time, Whereness offers the prospect of much more fine-grained advertising within indoor retail and leisure environments.

One of the original ideas of ubiquitous computing could be used to good effect. A public multimedia display can suddenly become personalized when a customer is automatically detected to be standing in front [7]. The advertisement, which perhaps first appeared on the mobile phone screen, is suddenly transferred to the public display. The mobile device can then become a remote controller to personalize an interactive session with the public screen (and perhaps a gamelike experience can ensue). There is a view that the future of shopping will converge with entertainment and Whereness could have an important role.

The market for location-based advertising would include organizations already operating on-line commerce and particularly those in sectors involving customers who are out and about, including:

- Shops and markets;
- Hotels and restaurants;
- Garages;
- Holiday and tourist facilities;
- Museums and outdoor cultural venues;
- Shopping mall owners and operators;
- Main street retailers;
- Sporting facilities;
- Airports and other transport hubs;
- Beaches and seaside areas.

The benefits are extended opportunities to target customers at the location of need or desire, where buying decisions will be enhanced by a desire to reduce stress and time wasted (often present when traveling or in unfamiliar surroundings). There could be a reduction in manual selling processes and increased automation concerning traditional printed or visual advertisements within facilities.

Weaknesses are, however, a potential resistance from customers if they feel the advertiser is intruding into their experiences or privacy. The reduction in scope for human to human interactions may impact negatively in some selling opportunities. Wireless technology may be subject to interference and poor performance in crowded indoor areas.

Positioning performance of around 10m accuracy with minute-by-minute updates is needed, with high security and low latency. There is huge potential to leverage the investments in RFID technology, currently used in the supply chain and for theft prevention, in a new way. Tagged objects on sale can react with customers and be part of an interactive experience. The tag readers embedded within the infrastructure could thus track customers and provide part of the Whereness service.

4.8 Personal Guidance

Guiding people is a service for people who are about to travel or who are traveling and also for people visiting a facility. It has synergy with the applications in ITS but is more general since people may or may not be using vehicles. ITS tends to focus on vehicles rather than the people in them. A wide range of devices can be used to interface the systems, which would be generally portable (with the exception of [personalized] public displays).

Before a journey is started, some interactions are necessary. For example, choosing a destination and selecting important parameters such as time, cost, modes, carbon emissions, and route. Confirmation can be made by the system, taking into account real-time system knowledge, such as delays or breakdowns. During the journey, a number of interactions with the central service can take place and include:

- Changes in a pre-calculated route due to new event information;
- Changes in a route based on new user destinations (i.e., an old journey is terminated early by the user and then a new journey is started);
- Progress information passed by the user to the central service so that the route plan can be checked regarding any time-critical issues (e.g., being late for something);
- Comfort information being passed to the user to update any time of arrival estimates;
- Emergency messages to alert the user of major events that might be important regarding changes of plans;
- Emergency messages involving messages being passed to the central service by the user requesting help and advice;
- At specific locations, the user may need associated secondary services such as permissions to enter facilities, transactions to pay for services, or commands to operate systems.

Post-trip information does not need to be in real time. It can be about performance information for planning and audit purposes, about payment information for budgetary purposes, and about health and safety issues.

Other services could take the form of automatic real-time booking or reservation services and messaging to update third parties about journey progress and delays. The markets would be for professional and business people who like to be in control or at the other extreme, tourists who want to be lazy about logistics. Specific groups would be:

- Field workers operating in distributed teams;
- Young adults and higher net worth older people with an active social life outside the home;
- Low-income people (without cars) traveling at public expense;
- Physically challenged people who need help while traveling;
- People who want to minimize their carbon footprints (or any other costs).

Benefits of personal guidance are save timed and other resources, better control of health and safety, reduction in temporal uncertainty, and better logistics.

The use of public transportation may increase as the experience is planned and monitored in real time on a personal basis. Road efficiency may increase, personal stress may be reduced, and scarce resources saved.

Weaknesses could be the lack of ability or desire by some people to follow machine instructions. This process can also be distracting and potentially dangerous, if appropriate media and displays are not used and if the user is not trained properly. Users may not trust the guidance being given and may believe it is being deliberately degraded (to the benefit of others) or for other malicious purposes.

Messages (or displays) offering guidance information must be timed early enough to be useful and not too late to be useless. Depending on the velocity of the user, the system would need a time window measured in seconds. Tracking accuracy can greatly be improved by map matching (assuming a suitable map with appropriate attributes is available). Choices of transport mode switch imply that information spaces from many different systems may need to be converged or mashed up. Some will involve static information such as maps and others dynamic information such as bus positions. Routes are usually well-trodden paths, therefore, some very simple positioning methods can be used in addition to the more sophisticated methods such as GPS. For example, a simple time delay or the reading from a pedometer or compass may be enough in some circumstances. Minimization of communications can be achieved along an agreed route that has been chosen in advance. Only when the initial plan experiences a deviation (due to a wrong turn by the user or an unexpected delay detected by the central system) does either party need to communicate [7].

Requirements include the convergence of all available portable positioning methods described in Chapters 6 and 7. Links to associated information spaces include GIS (for maps), identity management, security including DRM for images and proprietary information, timetables, public transport tracking systems, police, and security services. The route-planning software could include optimization, including facilities to monitor progress of each journey (both locally and centrally) in real time and include calculation of least time, distance, price, energy, and environmental carbon emission costs.

4.9 Finding People

Finding people is a service to find the real-time whereabouts of work colleagues, or family and friends. Users would be able to interrogate a database via a wide range of devices including the Web, SMS text, mobile phone, and fixed phone. They would then receive an answer to the question, "Where is <Name>?" Interfaces can be voice, textual, or graphical via a map and be in the form of a symbolic description such as an address or room number a relative distance away and some context information, for example, whether the person is driving, walking, or stationary.

Secondary services could take the form of audit trails, such as where the person was located relative to a time horizon, for example, during the last 2 hours. These could also include group applications where the service is extended to several people either identified individually or via a group identity. Knowledge of a relative distance (away) will imply that the systems also have knowledge of the users' location.

The main markets include professional business people who are working in urban areas and traveling by car or public transport, field workers operating in distributed teams, and young adults and higher net worth older people who have active social lives.

Benefits to users would be saved time and other resources by eliminating fruitless searching and avoidance of unnecessary time contingencies. Emotional stress may be reduced as uncertainty is reduced, and efficiency may be increased as work and leisure become more flexible.

Weaknesses of the service would be concerns about privacy and trust, so that many people may not want to participate as targets, and some may find ways of cheating. Good forward planning of meetings and events may be reduced, as reliance on dynamic finding of people leads to improvements for those with the service at the expense of those without.

Requirements are that targets should be located to a specific room, building, or street segment (preferably) within sight of the user's arrival, after following location advice. Location to within <10m is required in general. In some circumstances this accuracy may not be achievable, in which case even with reduced performance the service may still be useful. For example, if someone's diary indicated being in a certain town, a single communications session in that town will increase the likelihood of the target being where the diary suggests.

Convergence will be needed with all available portable positioning methods and close integration with electronic diaries and calendars will be desirable since this is important (a priori) information. The automatic service is then acting mostly to confirm the validity of a diary entry.

Associated information spaces include GIS (maps), identity management, and security including DRM for images, and proprietary and private information.

4.10 Moving Object Management

Goods within supply chains, vehicles carrying goods or people, fleets of vehicles in general, couriers, and mobile workforces all have a requirement to be managed efficiently. Workforce and vehicle utilization must generally be maximized and time between productive activities minimized. Allocation of resources needs to include the real-time management of what (or who) is where and when. Exceptions need to be generated if plans are not being fulfilled and reports made to controllers so corrective action can be taken, which may then include guidance services.

The markets for this class of logistics service include supply chain managers, warehousing, stock control, and delivery. Essentially, the application concerns the checking of goods out of one place and into another, either by reference to each item or by reference to the container of the item. There is a hierarchy where the goods may be individually packaged. The packages are grouped into larger boxes, boxes placed on pallets, pallets into containers, and containers hauled by trucks that may finally be moved by ship. Tracking the ship effectively tracks everything on the ship, if a database has recorded all the associations. Therefore it is very important in supply chain management to accurately record when one class of objects is associated physically with another. Clearly, a major reduction in human error can be achieved if the process can be automated. RFID tags, barcodes, and communications links all have an important role.

Freight, transport, and logistics operators can check the progress of loads in transit. Customs, security, and law enforcement agencies can monitor vehicles and their loads, crossing borders and other important boundaries (e.g., prisons, military bases, secret facilities, and schools).

Work managers can keep track of work by location and status of the vehicles used by workforces. Fleet managers can ensure that vehicles are being operated within the law and within organizational rules such as those concerning driving hours, speeds, and parking. The community as a whole can benefit if the vehicle positioning system is also used to monitor traffic delays and automatically feed back the delays to road traffic management systems.

Benefits include automation of record keeping and the elimination of human error, information availability in the event of natural disasters and terrorism, vehicle efficiency management, driver monitoring for health and safety reasons, load monitoring, and the general collection of positional evidence in case of disputes (particularly if the evidence is managed by a trusted third party).

Weaknesses include workforce resistance to micromanagement; a target for terrorists, criminals, and hackers, the relatively high costs of mobile communications, poor mobile network coverage and performance, especially in busy periods; incomplete mobile network coverage in rural areas; and the physical difficulty of provisioning power and antennae, especially on stacked metal containers.

Location updates are not necessarily needed regularly but when they are, updates are needed in numbers of minutes. The alternative is to send location and status information by event (i.e., at key stages of a journey), detected by location sensors or timers.

Many technologies may be needed but for vehicles, the use of GNSS is most likely. For indoor situations, the use of bar codes, RFID, and other sensors such as cameras are likely. Good-quality mobile communications such as GSM GPRS[5] are needed for vehicles or the use of mobile satellite services.[6]

[5] Global System of Mobile Communications (GSM), General Packet Radio Service (GPRS).

[6] For example, from Inmarsat, which has global satellite coverage.

This application involves event management, an intelligent function that associates a positional movement at various times (i.e., an event. See Section 8.8.5). Intelligence within the system involves following a set of given rules to check for problems. Integration with enterprise systems is likely such as stock-level management and work-flow systems. For example, if an object or vehicle is detected as late (according to a preset rule) in a specific place (such as a way point set by a geo-fence), a warning may be sent to a controller. The intelligence may be local, central, or both. Integration with a GIS is likely so that operations can be visualized as a graphical map.

4.11 Mobile Maps with GPS Overlay

One of the most widely used services on the Web today is the map and aerial photography services that are now available and that are gaining extra functionality such as the ability to include personal information overlays. Maps and GIS will be discussed in more depth in Chapter 8, but the basic application will be discussed here first.

Not only can personal data be included, but dynamic data such as personal positions as determined by a GPS receiver (or by other means such as the reception of a known WiFi hotspot) as well. Along with the vehicle sat-nav in car equipment, the Web map is perhaps the closest people have today to a real Whereness system.

The benefits are the scope of the mapping and imaging that already are global. In some dense urban areas and popular destinations, the images and maps are now 3D models. It is becoming possible to experience anywhere on Earth (and some other planets) by these geographic applications. APIs are exposed to both amateurs (for free) and to business partners (by commercial arrangement) so that integration with other services is now possible. Another benefit is of free service or rather one that is paid for by advertising. The use of mobile data access in the form of WiFi hotspot and cellular radio (2.5G and 3G) is gradually making an impact so that these maps and images can be used on the move, using either ultra-portable PCs or smart phones that now seem to be eclipsing the personal digital assistant (PDA).

There are many weaknesses but we can expect some of these to be rapidly addressed. As demand rises the quality of the imaging will improve, especially as imaging is improving rapidly as more spacecraft and surveillance aircraft are used. Real-time information may be included if the operators of the services become Whereness operators. The maps currently are only of outdoor locations, but maps need to include indoors and even extend underground and deal with topology in general.

The requirement to use these services in a Whereness context is the availability of positioning information. The accuracy is not a specific issue since it depends upon the use of the visual map or aerial image. For example, accurate

GPS tracks are needed to follow the progress of a delivery. If, however, I only want to let my friends know that I am generally "around" (i.e., not in another city but am unwilling to disclose exactly where I might be), very coarse accuracy might be helpful as, for example, provided by the local cellular base station using Cell ID. The maps could be overlaid in this case by a shaded or fuzzy area rather than a pinpoint icon.

4.12 Summary

This chapter considered the Whereness applications of today, particularly those that have been associated with ITS and researched in the 1980s and early 1990s. Mobile information was discussed including the RDS-TMC system that uses a simple location code. A more advanced application is dynamic route guidance that involves a set of interrelated databases necessary to create useful routes factoring in future traffic congestion predictions.

Tolling or the collection of money with respect to position was discussed and also the accounting of other value such as calories or carbon emissions. Parking, ticketing, emergency calling, tracking, and logistics were other ITS applications discussed.

More general location-based services described included mobile advertising, personal guidance, finding people, and the management of moving objects. Finally the importance of maps and Web 2.0 was again highlighted.

References

[1] *ERTICO ITS Europe*, http://www.ertico.com/en/activities/efficiency__environment/road_traffic_information_group_.htm, Jan. 2008.

[2] *Ali-Scout*, http://www.umich.edu/~driving/publications/UMTRI-96-32A3.pdf, Jan. 2008.

[3] Catling, I., et al., "SOCRATES: System of Cellular Radio for Traffic Efficiency and Safety," Proc. of the DRIVE Conference, Brussels, Feb. 4-6, 1991.

[4] Golding, D., et al., "Mobile Multimedia Applications," *British Telecommunications Engineering*, Vol 17, Part 1, Apr. 1998, pp. 20-21.

[5] Randall, P., et al., "PROMISE A Personal Mobile Traveller and Traffic Information Service," 4th World Congress on Intelligent Transportation Systems, Berlin, Oct. 21-24, 1997.

[6] Mannings, R., Wall, N. D. C., Navigation Information System, European Patent EP0777863, U.S. Patent 66664924, first filing 1995.

[7] Patel, D., Mannings, R., "Reflex-Personalised Wireless Interaction in a Broadband Environment, 2004," *BT Technology Journal*, Vol. 20, No. 1, Jan. 2002.

Chapter 5

Future Whereness Applications

5.1 Society Futures

Before considering more futuristic uses for Whereness that could be used when it is a mature technology (perhaps by 2015 or earlier in the cases of some component parts), it is important to consider how societal and business environments might have changed by that time. There is a growing trend in business to actively study the future:[1] some publications are becoming available that inform the activity, for example, the BT Technology Timeline [1]. The following sections apply current corporate foresight thinking to the future of Whereness.

5.1.1 More Wealth

Figure 5.1 is a version of Maslow's hierarchy of needs [2], which represents resources spent by society as a set of layers. The size of the layer represents the importance and cost of the various needs that are ordered in terms of criticality. At the lowest layer are the essentials for survival. In the middle layers are the important future needs but not strictly essential needs; examples include education and family relationships. At the very top, where resources are thin, there is a small residue of resource for "self-actualization" that would include, for example, participation in the arts.

[1] Future study is often referred to as "foresight" or "futurology." It is a multidiscipline activity and includes economics, sociology, politics, and science/technology.

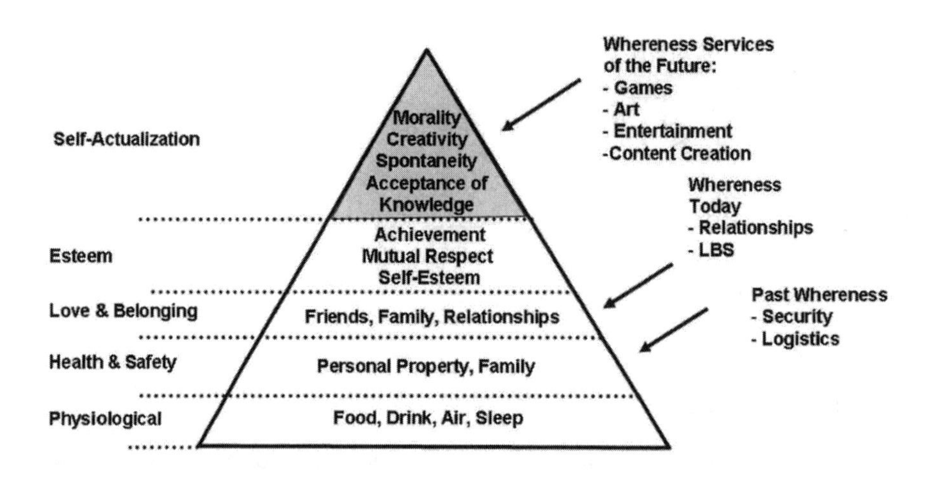

Figure 5.1 Maslow's hierarchy of needs applied to Whereness.

It can be argued that as the developed world gets steadily wealthier, as it has since the Second World War ended, the pyramid is inverting in relative terms. We still need the essentials but more and more wealth is focused on self-actualization, which generally are things and activities that appeal to our emotions. In economic terms, the utility function of many things now includes increasing emotional value in addition to material value.

Today we are seeing Whereness beginning to be applied to the middle layers with LBS and social messaging based on location, but in the past the areas of Whereness were mostly niche and applied to security and logistics. In a narrow sense these are more useful but as wealth increases useful is being redefined.

5.1.2 Neglect of the Basics

One consequence is the neglect of the lower layers of the hierarchy as politicians, business people, and the public focus on the newer "essentials." A changing climate is one of a number of trends that could highlight this neglect, forcing governments, in particular, to spend more on environmental monitoring and disaster response. Whereness is likely to be essential to help manage the logistics of a more uncertain world.

5.1.3 Increased Automation

When disasters occur there is often a chaotic response due to information overload and staff who are not trained in complex software systems such as GIS. The author discusses this topic in [3] and suggests that more information automation is needed. GIS, for example, needs to "disappear" with simple instructions issued by the machines to help solve problems quickly during disasters.

5.1.4 Human to Human Technology and the Digital Bubble

If Whereness is going to help automate the lower layers of Figure 5.1, what of the uppermost layers? What can Whereness do regarding human emotions? The other major role for technology is to help increase the human aspects by enabling rich new social, artistic, and entertaining experiences.

A powerful idea for the future is a "digital bubble" of ambient intelligence that follows people around and also interacts with the bubble belonging to other people and physical artifacts. This idea is explored by the author and his colleagues [4] using the metaphor of "virtual air." For example, when two people meet (and their personal profiles are set to allow an exchange of information between the bubbles), "magic" can be made to happen. Other applications include dating services for the unattached, fun and games as one enters a pub, or a service to guide a group of people to somewhere mutually beneficial.

5.1.5 Big Brother

At the other extreme of the hierarchy is the most important of the needs, that of security. If the threat from asymmetric warfare (which is perhaps a more accurate term than terrorism) increases, the issue of more tagging and tracking the general public will be raised more frequently. Depending on how bad the threat becomes, particularly if the threat starts to involve smart weaponry or unconventional attacks, the democratic process will need to decide on the trade-offs between privacy and security.

5.2 Health and Well-being

5.2.1 Calorie Counting

Obesity is one of the major challenges for Western society. Although some people are living longer we are now seeing a reversal of this trend for others. To promote and maintain fitness, there will be automatic services to record and manage the daily exercise taken by people based on their physical (pedestrian) movements. It will include the positional information concerning gradients and contexts such as walking up stairs (or alternatives such as elevators that might be highlighted as undesirable). Automation will mean the user does not have to do anything other than carry appropriate technology, which will be, in due course, integrated within clothing. The recording of movement will be linked to an individual health and fitness plan that would be incorporated into the personal profile. The active part of the service could take the form of an artificial coach that encourages the participant to exercise properly.

Markets for individuals would include health-conscious young professionals, middle-aged people with a need to lose weight and improve cardiovascular condition, active seniors, and sports enthusiasts. Business customers would include organizations with workforces keen to improve their well-being and to avoid future litigation, education authorities promoting a healthy lifestyle, life insurance companies as part of a strategy to reduce risk, and national public health services.

Benefits would be a reduction in the costs of future healthcare, improving fitness, and personal well-being. Transport costs for individuals would be less by promoting walking and running. Longer life spans may result and increase of life quality when aged.

Weaknesses, however, could be that health benefits may be found to be false or lead to unintended results (e.g., increased joint wear). People may reject the technology because they feel under pressure to conform to an imposed fitness regime. An increase of pedestrian accidents may result.

Around 10m accuracy is likely to be adequate but measurements need to be continuous to detect velocity. Vertical positioning accuracy of 3m is desirable. Networked pedometers are needed and other sensors also including accelerometers, altimeters, and gyroscopes so that other measurements can be taken to detect vertical motion and general position (see Section 7.5). The system need not work in real time, which could result in no requirement for wide area mobile data. Indoor wireless hotspot communications are likely to be adequate to collect statistics and download maps and fitness schedules. Integration with other lifestyle software and systems is likely, including the automatic counting of food energy (calorie counting), the control of gym equipment, sporting performance monitoring, life insurance monitoring, and workplace health systems.

5.2.2 Caring for People (and Animals)

The care of people who have special needs or are vulnerable can be augmented by the use of positioning technology either as a primary aid to mobility or as a method to increase their freedom while ensuring their well-being and safety. For example, a blind person with a guide dog may use a positioning system to help identify building numbers or public transportation [5]. Children may be given

more freedom to roam if their caregivers know where they are at all times. The same approach can also be adopted for the care of animals.
There are two approaches. First, an active approach, whereby the position of the person is determined by equipment carried by him or her, which then communicates with the infrastructure. The second approach is passive, where sensing systems are used to recognize movements, usually within a bounded area. The advantage of active systems is that they are high-performance and can be used to follow movements anywhere, but it requires a Whereness device to be carried that may be bulky, heavy, expensive, prone to failure, and requires a working battery. The passive system, in contrast, requires no equipment but will only work in certain areas covered by the fixed sensing system (e.g., surveillance cameras, floor pressure sensors, passive infrared systems, and volumetric ultrasonic systems). A second problem with the passive approach is the reliable identification of the target. If there is more than one person being tracked, a secondary system is needed to identify them.

Markets for this service would be mostly the caregiving agencies rather than the individuals in care. These would include:

- Blind people;
- Deaf people;
- People who are physically or mentally challenged;
- Children in care;
- Vulnerable workers in difficult areas;
- Prisoners in the community;
- Pet management;
- Farm animals;
- Endangered species in the wild.

Benefits are a reduction in the cost of care since automation gives caregivers more freedom and flexibility, which can decrease stress of users and caregivers.

Weaknesses, however, can come from abuse by criminals or poorly performing caregivers. Exploitation by authorities is also possible, eliminating people from areas where they are still needed. An overreliance on radio (always subject to potential interference), poor battery performance in the active case, and radio coverage are all areas that need to be carefully managed.

Performance requirements are exacting. High levels of positioning accuracy are needed (<10m in general) and very low latency. Very high levels of security are required. Tracking needs to be continuous, with positions recorded second by second. There may be a need to extend coverage indoors using sensor technology. This is probably the most demanding of all Whereness applications.

Very good quality mobile data communications outdoors is needed, with possible need for roaming between networks. A GIS is required to manage geo-fences either centrally or locally to raise alarms when targets stray. Intelligent software functions are needed to associate position with information for each user and appropriate human machine interfaces (HMI), particularly for those with special needs (e.g., haptic technology or technology using the sense of touch for blind people). Figure 5.2 shows a blind person exploring a digital map using a haptic interface, and Figure 5.3 shows a blind volunteer using a personal guidance system with DGPS positioning and a digital map attributed with spoken voice information.

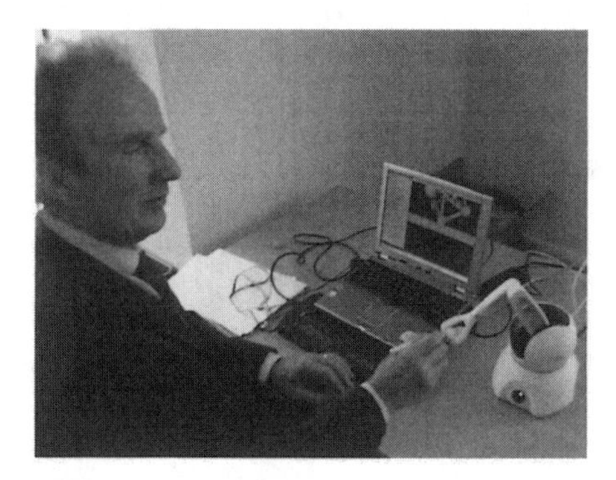

Figure 5.2 A blind volunteer researcher using a haptic map interface to explore a model of the Brussels' Atomium building.[2]

Figure 5.3 A blind volunteer using differential GPS personal guidance system[3].

[2] Project ENABLE, E.U. 6th Framework Programme.
[3] Project MoBIC, E.U. TIDE Research Programme.

5.3 Floods and Disasters

5.3.1 Sensor Networks and Key Asset Management

Sensor networks can monitor the position and status of key assets such as search and rescue vehicles and people, the state of water levels and flooding (using moving floating sensors), and the position of flood waters. These applications are likely to increase as climate changes. There are predictions for more variable weather that will lead to an increased requirement to manage the ensuing problems.

5.3.2 Real-Time Maps, Imaging, and Photographs

In addition to sharing my whereabouts track with the government, who else might I share it with? Would I be prepared to share a favorite walk or a new ramble for the benefit of strangers who share my love of the countryside? This sort of service is very close in concept to the sharing that is taking place in social networking Web sites. Whereness can then extend the social Internet into the real world on the move. Maps, photographs, and any other multimedia can have both a real-time temporal and spatial aspect. Clearly, trust is important and nobody would want vulnerable people who are alone being followed by potential abusers.

The concept of people building and sharing maps (with media embedded) will be discussed in detail in Chapter 8, but the content of the Internet has only grown by the paradigm of sharing and it is important to continue this movement but with appropriate safeguards.

5.4 Countering Terrorists

It is likely that the threat from terrorists and those engaged in asymmetric warfare will increase. So far most attacks have come from the use of very basic weaponry but that situation may not last. Increasing availability and capability of advanced wireless, sensors, and computing devices may lead to the creation of new generations of smart booby traps, roadside bombs, and, conceivably, guided weapons.

Although it may be possible to deny access to some components and systems, it is still possible to hack consumer devices and adapt them for terrorism.

Whereness technology would be especially useful to guide and deliver explosives, chemical poisons, and biological pathogens.

However, Whereness can be used as a countermeasure on both passive and active tracking systems. In particular the transport networks can automate the ability to monitor huge numbers of potential suspects. It is also possible to highlight unusual movements of people and objects to identify new suspects using massive deployments of surveillance cameras with pattern recognition and data mining.

5.4.1 Big Brother Tamed?

Perhaps a distributed approach to Whereness (which is the main recurring theme in this book) can help by encouraging transparency. For example, consider this hypothetical conversation by a law-abiding citizen: "I might be happy for the authorities to know where I have been, if I know what information they actually have on me. How do I know? It came from my own portable system, so I can check and if necessary show someone the master log file." The main worry is not so much law-abiding citizens being concerned by being tracked, per se, but the prospect of being falsely accused by a central system that is either incompetent or that has been hacked in some way.

If Big Brother is required to maintain the operation of a (reasonably) free society, then there is a growing onus on the authorities and their partners to assure the integrity of systems. There is a corollary, perhaps, with the credit-checking agencies. If people suspect there may be a problem with their records, they have a right (in some countries at least) to access the records and have them corrected. Perhaps a similar situation could emerge for mass tracking where the trusted agency can allow legitimate requests from people to check whether their whereabouts are accurate. The use of personal blackboxes (discussed in Section 9.2) might mean that individuals have much better quality of information than the authorities.

5.5 Sports and Games

New generations of location-aware games and sports are being researched. Sports where the participants are not colocated are possible with a sensing system providing positioning information of players, balls, and equipment. Breakout for Two [6] was ball game where two remote players used mutually connected video walls to play a ball game, where the ball's bouncing position was sensed locally and relayed as an image to the screen at the remote end.

5.6 Mapping Everywhere

Simultaneous location and mapping (SLAM), which will be discussed in more detail in Chapter 9, is a future Whereness application that comes as a spin-off from guidance applications in mobile robotics. It is a positional information scavenging system that collects potential map information from positioning measurements and thus builds maps, while the main system is simultaneously providing the positional information to more conventional applications such as guidance. Since these systems can be greatly improved by using map matching, the repeated use of SLAM builds, improves, and uses maps simultaneously. The value of the map is increased and it is maintained automatically. It can discover new areas to be mapped and highlight changes to existing maps. A validation method (perhaps provided by an agency) is required to accept any changes before they become persistent.

The markets for SLAM are mapmakers, GIS organizations, and map users. Map validation service providers would sell the maps that have been checked to many current GIS and map users including building and facilities managers, utilities, the construction industry, local authorities, and governmental agencies. All these users would most likely be in partnership with the more general Whereness service providers.

Benefits of the application are the creation of free maps as a by-product of other commercial applications. Map attribution can be automated if features can be identified using software, and map maintenance can be improved by changes being automatically highlighted. Existing mapmakers may have to lower prices if SLAM is taken up by open-source mapmakers universally.

The weaknesses of SLAM are the users must be persuaded to participate since there are no direct benefits unless incentives are included in some way. Some map attribution is still needed after map creation and certainly validation would be necessary. In some areas SLAM may be forbidden for security or legal reasons. Hackers and criminals may seek to create false maps.

SLAM has an exacting requirement for accuracy and coverage (the higher the better). Galileo, with <1m accuracy, will be very useful. There is no specific need of real-time communications between mobile mapmakers and central systems. The systems will collect map fragments and meld into a tessellating whole GIS. It is unlikely that this process would ever be entirely automatic because of ambiguities. (See details on open mapping in Section 8.7 and SLAM in Section 9.4).

5.7 Locative Media

Locative media is a very important part of the future electronic media. By using positioning technology to trigger the delivery of multimedia in particular places, engaging experiences either individually or in groups can take place. For example, a narrative could be created that is delivered by people wandering at will around an area. The exact story would vary and interactions between strangers could form part of the experience [7]. This new form of media interaction has relevance to the future of entertainment, sports, games, and art.

5.7.1 Conceptual Art

Conceptual art may not be a familiar term and its relevance to Whereness may not seem obvious as art and technology may appear to be very separate areas. It can be argued, however, there has always been a close symbiosis and new art movements using technology such as conceptual art are nothing new [8].

Conceptual art includes the idea that the artist creates a machine that creates the art. In particular, the audience or observers can use a situation provided by the artist to engage in an activity that is a work of art. This set of applications somewhat defies description or rather prescription (what constitutes art is unclear). Artists will always take new tools and find novel ways to use them. The tools provided by ubiquitous positioning will enable experiences in one place to be reflected in other places and vice versa. Art can be everywhere, and everywhere people are, art can be experienced via electronic media. The spatial-temporal context can be communicated and then coupled with virtually any other medium. Mobile participants can experience the art or even be the art. The normal 4D world will merge with other virtual worlds. In the much longer term, from a transhuman perspective, even the artist and audience could be partially nonhuman.

This is the most demanding set of applications. Wireless technology is not always adequate so reliance on video capture of motion is needed (see Figure 7.3). Groups of voxels[4] are handled in what is effectively a four-dimensional GIS (i.e., 3D space plus time). Integration with many types of media including touch technology or haptics is possible. Thus an object that is somewhere remote can be experienced locally in a very tangible manner.

There is a requirement for nonintrusive technology, where people do not need to wear or do anything technological (i.e., using field sensing and video recognition of the body and its motion). In the future there will also be a need for integration with robotics, actuators, biometrics, and longer-term coupling to the brain and nervous system.

5.7.2 Augmented Reality (AR)

Augmented reality (AR) is related to the better-known concept of virtual reality. Virtual reality (VR) concerns activities in artificial worlds entered via a computer's human machine interface (which are usually part of a gaming or social experience). In contrast, augmented reality concerns activities in the real world

[4] A voxel is the three-dimensional equivalent of the two-dimensional pixel. Most computer-generated screen images are composed of pixel arrays.

with content and interaction from artificial worlds overlaid. This is best explained by an example. Suppose a group of friends were playing a game involving chasing software-generated monsters. They could be running around in a public place outdoors but would see and hear (and potentially feel) the monsters using a variety of multimedia devices, some of which would be carried and have screens (along the lines of a camera viewfinder or even a head-up display) and some of which could be (multipurpose) public displays. Some examples of some recent research experiments are given in [9, 10].

Whereness is a key component of these systems since the media is locative (i.e., connected with a location). The exact temporal-spatial positions of the users in the real world are linked very closely with their equivalents in the virtual world or worlds.

Augmented reality is a very important aspect of the future in many ways. First, the experience is likely to be highly engaging and entertaining in a human way and thus likely to be both popular and lucrative commercially in common with most computer gaming (now larger than the movie industry). Second, the technique suggests that perhaps we should start to think of virtual worlds as extra dimensions of the real world. Rather than concentrate on totally artificial world creations, some of which have raised concerns about the psychological dangers of decoupling from reality and thus by extension from the norms of morality, perhaps it would be better to add extra layers and dimensions to the real world? An example might be the apparent ability for quasi-time travel. For example, here is a scenario that might appeal to a student of history.

Ten kilometers from the author's home is the famous Anglo-Saxon ship burial site of Sutton Hoo, where it is thought one of the first kings of England was buried with his treasure, in a ship covered by an earth mound near a river bank. The only traces of the ship found by the archeologists were the iron rivets and the impression in the soil of the wooden planking that had rotted away during the past 1400 years. It would be an amazing educational and cultural experience if we could now recreate an augmented reality version of the ship and site, where one could walk around and see (via a personal display) what it would have been like in 600 AD. For example, see the (software-generated) ship sailing on the real river and be ushered into the presence of the (virtual) sovereign. Although there is a modest real tourist experience today, the costs of providing the augmented reality version would also be modest in comparison with a major theme park that would be unlikely to be funded in this example.

The technical challenges of providing vector graphic VR models that are linked in real time to personal positioning systems of high accuracy are only just becoming feasible, but it is likely that in a few years' time the technique will be commonplace.

5.7.3 Augmented Reality Support

Services supporting AR within an area, must provide accurate positioning information to facilitate the display/capture of locative media in appropriate places. A variety of HMIs cab be used including wearable consumer head-up displays (HUDs), handheld small screens and linked public displays, all of which react in response to user motion and context. Markets include:

- AR games and especially collaborative gaming;
- Collaborative and distributed sports (e.g., geographically separated racing);
- Educational overlays for out-of-school activities;
- Multimedia narratives for tourism;
- Enhanced experiences for museums and historic monuments both indoors and outdoors (e.g., showing original historical context and in-filling of missing architectural artifacts);
- Virtual tourism;
- Cinematic activities, blending real object movement with virtual object movement.

Benefits are enhanced social, educational, recreational, and cultural experiences that may be highly engaging, popular, and thus commercially attractive. There may be cultural benefits leading to opportunities to engage with people in other regions by "making geography, history," including travel reduction leading to cost savings and reduction in environmental damage.

Weaknesses include more public backlash about artificial experiences, and there may be possible safety hazards (e.g., walking into things) and dangers of new social experiences and norms, leading to unacceptable behavior.

As well, very high accuracy location would be needed in certain locations (<1m) and there is a need for accurate orientation. Very low latency is required; for motion picture capture accuracy <10cm is required.

This is a very demanding set of applications. Wireless sensing technology is not always adequate so reliance on video capture of motion is needed, as explained in Section 7.4.5. Integration is needed with 4D GIS (i.e., 3D space plus time—See Section 8.8) and integration with many types of media, including haptic or touch technology.

5.7.4 Scenarios for the Future

The end of Chapter 9 discusses a Whereness vision in terms of user equipment and services. It continues with the themes in this chapter, which are futuristic and will mostly not be common for at least a decade. The next three chapters are more concrete and focus on the key technologies of Whereness today and into the medium term. Chapter 6 concerns radio positioning, Chapter 7 nonradio sensing systems, and Chapter 8 is about maps, mapping, and the very important current developments in Web 2.0.

5.8 Summary

This chapter explored the future application for Whereness but started with a discussion of the future of society as informed by futurology. The increase in wealth and movement up Maslow's hierarchy of needs will lead to more emphasis on applications concerning emotions, self-actualization, art, and entertainment. There is a danger, however, that the basics of sound infrastructures and the needs for good security may be neglected. Whereness applications are important for both extremes of need, and adoption of Whereness for social uses may help the case for security applications.

Several future application scenarios were explored. Health and well-being included the need to automatically count calories and care automatically for people and animals, especially people with special needs who are vulnerable. Floods and disasters will probably become more common, so sensing applications are important to both monitor and manage the environment and counter terrorism. Sports and games will evolve new formats where positioning systems can aid the linkage of remote environments. New maps will be built and managed automatically by people sharing positional information. Locative media and augmented reality, which are still emerging technologies, will become common and add exciting new experiences ranging from gaming and entertainment to new forms of conceptual art.

References

[1] Pearson, I., Neild, I., *BT Technology Timeline*, http://www.btplc.com/thegroup/Publicaffairs/EuropeanAffairs/Thetechnologytimeline/timeline.htm.

[2] Maslow A. H., "A Theory of Human Motivation," *Psychological Review*, Vol. 50, 1943, pp. 370-396.

[3] Mannings, R., Parker, C., "The Invisible GIS: Technology Convergence to Make the Future GI User Friendly," *AGI 2006*, London, Sep. 2006.

[4] Mannings, R., Perason, I., "Virtual Air," *The Journal of the Communications Network*, Vol. 2, Part 1, Jan.-Mar. 2003, pp. 29-33.

[5] Pinnick, T., *Enabled-Enhancing Network Access for the Blind and Visually Impaired*, http://www.hft.ogr/HFT06/paper06/11_Pennick.pdf, Jan. 2008.

[6] Mueller, F., Agamanolis, S., http://web.media.mit.edu/~stefan/hc/projects/breakoutfortwo/, Jan. 2008.

[7] Strohecker, C., *Nature Trailer*, http://www.carolstrohecker.info/PapersByYear/2006/NatureTrailerBrief.pdf.

[8] Mannings, R., "Everywhere Art and Art Everywhere," *The Journal of the Communications Network*, Vol. 5 Part 4, Oct.-Dec. 2006, ISSN 1477-4739.

[9] Bulman, J, et al., "Mixed-Reality Applications in Urban Environments," *Intelligent Spaces: The Application of Pervasive ICT*, Steventon, A., Wright, S., (eds.), London: Springer-Verlag 2006, pp. 109-124.

[10] Lancaster University Guide project, http://www.guide.lancs.ac.uk/overview.html.

Chapter 6

Radio Positioning

6.1 Radio Positioning Basics

This chapter and Chapter 7 will be considering the science and technology of positioning systems. All the principle methods were discussed briefly in Chapter 2, which included a time-line representation (Figure 2.1). It can be seen that some methods use radio systems and others use various other technologies that sense other physical phenomena. Radio will be discussed in this chapter, since different radio systems have many characteristics in common. It will be necessary to delve into some theory but the main focus will be to help the reader understand the characteristics and compromises. To reiterate the point made in Chapter 1, there is no perfect method for ubiquitous positioning, rather a convergence of systems and fusion of information.

Radio communications can appear to have a magic quality. Radio signals travel though solid walls (apparently) instantaneously, can travel very long distances, appear and disappear into and out of electronic equipment, and are the subject of military and covert activity. Systems can have an enormous price ticket yet are finding their way into more and more everyday objects and appliances. Radio is a very important aspect of ubiquitous computing and although radio will not replace fixed communications, increasingly the first (or access) part of networks is wireless. In spite of its increasing importance in business and society many people have little idea of how it works and thus what it might lead to. The intention in the next few sections is to explain why and how radio aids positioning and to expose a little of the underlying theory to help the reader understand the principles. There are many excellent books on radio theory but the subject is deeply mathematical and this can be a barrier. A more pragmatic approach is, however, taken by radio amateurs so it is recommended that their handbooks are consulted first by anyone looking for practical advice and information [1] and also a basic primer [2]. It is interesting to note that amateurs have made a huge impact on the advance of radio during the 20th century, in a way to which, more recently, similarly motivated people have pushed forward the Internet, the Web, and even more recently open mapping (covered in Chapter 8).

For up-to-date background information on digital wireless and mobility, the *BT Technology Journal* is an excellent resource that strikes a good balance between the needs of business people and those more technically inclined. Several special editions are relevant to wireless and Whereness, as follows. "Localisation and Personalization" [3] covers location-based services, locating calls for the emergency services, cellular positioning, and a general overview of positioning and personalization. "Telecoms Unplugged" [4] covers topics such as wireless LANs, ultrawideband, 3G and beyond, spectrum, and mobile multimedia. More recently "Mobility and Convergence" [5] covers some digital wireless fundamentals, ad hoc wireless, trust, convergence, various Internet topics, and WiMax.

The words wireless and radio seem interchangeable but this can lead to confusion. Optical communications are wireless but not usually thought of as a radio methodology. Radio is usually wireless except that the radio frequency (RF) electrical signals often pass along wires within equipment. Optical signals (i.e., infrared, visible light, and ultraviolet rays), and radio are all forms of electromagnetic radiation and are physically the same. We describe them as waves traveling at light speed (300m m/s) in a vacuum or somewhat less through any other transparent medium. The higher the frequency, the shorter the wavelength, which is measured in meters for domestic radio, millimeters for microwaves, but nanometers (billionths of meters) for optics, as shown in Figure 6.1.

Although some areas of physics require that radiation be described as particles (photons), for the purposes of positioning technology we can avoid that complication and consider waves alone. Waves normally propagate through a medium but the concept of the ether was abandoned as unnecessary. Essentially, radio waves are rapidly changing electrical and magnetic fields that react magnetically and electrically with materials, in particular with good conductors (i.e., metals that are usually used to construct antennae). A wave will start to propagate when a changing electrical or magnetic disturbance is set up in a transmitting antenna and will continue in a uniform direction in a vacuum without loss of intensity or power. If the wave encounters any material, it will react by inducing a similar disturbance to that which caused it. Antennae are made to resonate at specific wavelengths and to interface efficiently with electronic equipment so that as much power as possible is directed in the required direction. Positioning technologies that depend on the directionality of propagation therefore depend heavily on good antennae.

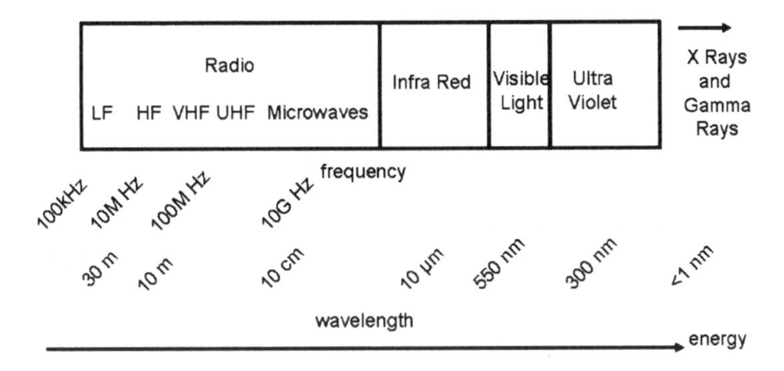

Figure 6.1 The electromagnetic spectrum.

6.1.1 Frequency, Wavelength, Bandwidth, Power, and Noise

Each color of light and each oscillating radio signal have a frequency f measured in cycles or oscillations per second for which the unit is Hz (hertz). It has a wavelength λ, which is related to light speed v, as shown in equation (6.1):

$$f = v / \lambda \tag{6.1}$$

In general, however, most radio systems operate over a band of frequencies, since when a single frequency is used to carry any information it is "modulated"—continuously modified in amplitude, frequency (or both). Modulation creates extra frequency components either side of the original "carrier" frequency. The more information that is sent, the greater space these extra frequencies (known as sidebands) occupy in the electromagnetic spectrum (see Figure 6.2). The frequency space occupied is called the bandwidth of the signal, and radio channels are allocated so that each channel occupies its own band without interfering with adjacent channels allocated to others. Bandwidth is a very important characteristic in positioning. The greater the bandwidth (in Hz), the more power (in Watts) is needed to transmit and the more accurate is any resulting timing measurement. There is always a limit to transmitter power because of power supply, heat generated, human safety, devices used, and especially radio regulations. However, the more power the better as far as the receiver is concerned.

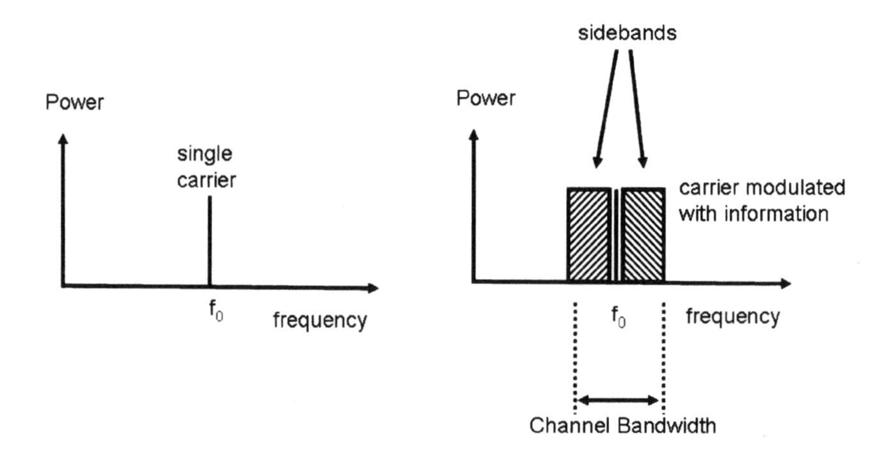

Figure 6.2 Bandwidth of a radio channel.

At a receiver in addition to receiving the desired signal (S), unwelcome extra signals are present including interference from other transmissions and an unavoidable component of random fluctuations present in all physical systems. Collectively, all these unwanted components are known as noise (N). Any system seeks to maximize the signal-to-noise ratio. The S/N ratio is purely a number but is usually expressed using the unit decibel or dB. The ranges in power in radio systems are very large, ranging from transmitters outputting kilowatts to receivers inputting picowatts (i.e., a range of 10^{15}). The decibel is a logarithmic ratio that helps keep the quantities manageable and allows gains and attenuations to be treated as simple additions rather than multiplications. The implication for any positioning system is profound: the better the signal-to-noise ratio (S/N), the more accurate any measurement is likely to be.

$$P_{dB} = 10 \log_{10}(P1/P2) \tag{6.2}$$

6.1.2 How Does Radio Provide Positions?

How does a radio system actually function and how can it provide a position? Firstly, the electronics in a transmitter generates a rapidly changing RF signal in the required frequency band and it is modulated with any required information at the appropriate power. The signal is carried by a "feeder" cable or waveguide (a hollow pipe used for microwave transmissions) from the transmitter to the antenna (or array of antennae) that is usually made of metal and shaped in special configurations. They are often tuned to resonate over a particular frequency band. The electromagnetic currents that circulate in the antenna cause electromagnetic radiation to be radiated into free space and the antenna configuration and shape greatly influence the efficiency (or gain) and directionality of the transmitted signal, as shown in Figure 6.3. It can be seen for equation (6.2) that the power gain of a typical parabolic antenna (such as would be used for a RADAR would be typically 40dB or 10,000 times).

At the receiver the opposite process occurs. The incident wave induces small electrical currents in the receiver antenna that are then conveyed by a feeder to the receiver, which is at its heart a powerful amplifier to restore the level of the signal to useful levels and a filter to select just the required band of frequencies from the host of unwanted signals usually present. The high directionality of both transmitter and receiver antennae is put to good use in positioning systems to measure angles with the highest gain antennae giving the most accurate angle of arrival readings. There is reciprocity in the way both (passive) transmit and receive antennae function and in many transceivers a common antenna is used. Active antennae, however, are used only in receivers and involve amplifier circuits integrated with the metal receiving conductor and do away with the need for an expensive low-loss signal feeder. They can be useful since they act over a very wide set of bandwidths.

Radio waves are also polarized, which is why, for example, some antennae appear horizontal and others vertical. The electromagnetic waves are thought of as having two components at right angles to each other: one electric and the other magnetic. In a conventional vertical dipole, for example, the electric field is vertical. Circular polarization is also possible, a hybrid of vertical and horizontal polarization, and is often used in mobile situations where constant alignment is difficult.

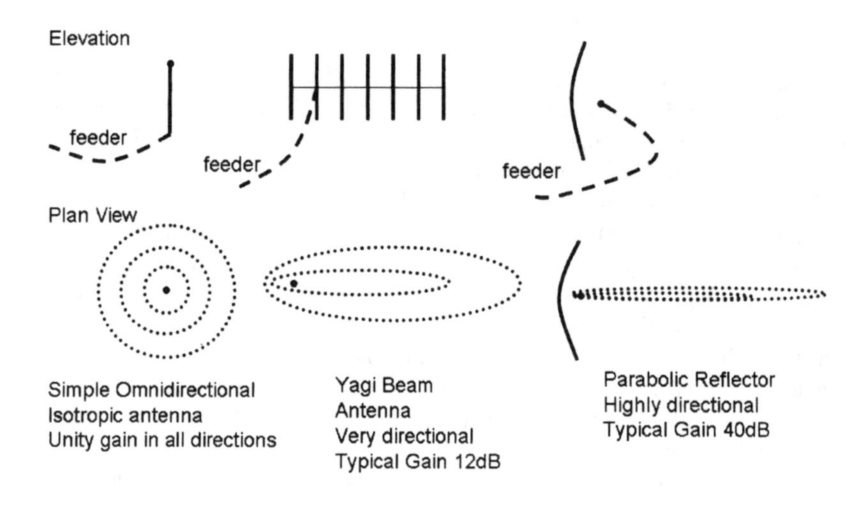

Figure 6.3 Simple common antennae configurations showing directionality.

To measure distances, the temporal aspects of the signal must be exploited. There are two methods used. First, and more commonly, the transit or flight time of a signal is measured. Many systems use narrow pulses or rely on the timing of various epochs in digital modulation (i.e., the start and end of a pattern of binary changes). From (6.1) it can be seen that in one microsecond the signal will travel 300m. The accuracy of the positioning system is then down to how accurately the signal can be measured and how good an S/N ratio is present. Noise can be thought of as power or voltage-level fluctuations but when these are passed through a threshold detector in the receiver that is a part of the demodulation process, they are translated into time jitters (see Figure 6.4). Further inaccuracy occurs when the timing is sampled by the receiver's digital processor since there is always a minimum sampling period usually determined by the clocking and logic speed of the processor's digital electronics. A fast processor may be able to sample to within a few nanoseconds, but this translates to 0.33m per nanosecond.

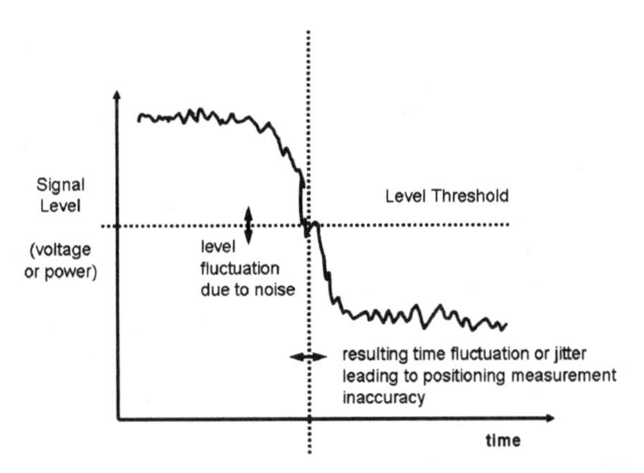

Figure 6.4 Positioning errors due to noise.

The second approach to temporal measurement is to use the phase of the carrier. A pure unmodulated signal is a sine wave with its phase varying between 0 and 360 degrees over each cycle. If such as signal is compared with time reference, a phase measurement can be taken. Thus two similar signals can be compared and the time shift expressed as a phase difference, as shown in Figure 6.5.

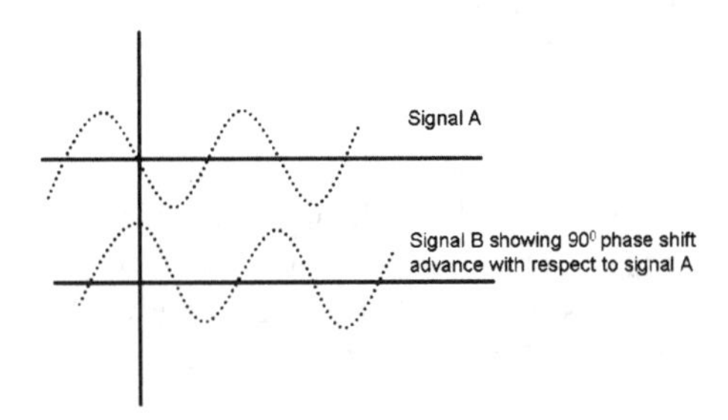

Figure 6.5 Phase shift between signals.

If a reference carrier is produced locally in the receiver and compared with the incoming signal, there will be a fixed offset between the actual cycles. For example, carriers would normally be pure sine waves that have regular zero-crossings so the zero-crossings of the reference and incoming signals could be timed. As the receiver moves the cycle offset will change with distance and repeat for each wavelength moved. Very accurate positioning can be achieved by this interference-like technique that has been developed from radio-astronomy. A more simplistic approach to radio positioning need not involve either the measurement of timing or angles but rather rely on the absence or presence of a signal when the receiver and transmitter are within a certain degree of proximity. Usually it is not strictly the true absence of signal but rather the fact that the power level of the signal has fallen below a level threshold set in the receiver. Many receivers contain received signal strength indication (or RSSI) circuits that give readings dependent on the level of the incoming signals. Sometimes these are displayed to the user and can aid manual frequency selection or tuning. The accuracy of the proximity measurement is usually very poor because of the wide variability of signal strengths in the environment. Nevertheless, it is a simple and cheap approach and is used widely since it needs no extra equipment for the positioning function. Usually antennae are used that have low directionality or are omnidirectional.

6.1.3 The Radio Propagation Environment

One of the most inconvenient aspects of radio does not concern the equipment at each end of the links but rather what happens to the signals in between. If our positioning systems were operating in deep space with a vacuum and no material between or anywhere near the antennae positioning would be idealized. In free space, the power density (p) of the signal falls as the square of the distance (r) between transmitter and receiver (6.3). It is thus possible to accurately map the radio-universe with highly directional radio telescopes. Unfortunately terrestrial signals are subjected to a number of disturbing terrain factors.

$$p \propto 1/r^2 \tag{6.3}$$

Although in general signals travel in straight lines, being waves they can diffract and reflect off surfaces and edges. This can be an advantage for mobile communications, since in dense urban situations there is very rarely a free line of sight between transmitter and receiver. Coverage in buildings, for example, is made up of multiple reflections and refractions. For positioning this is very unhelpful since the line of sight is the shortest path and it is mostly absent, with a whole host of signal fragments coming from many directions with differing flight times. Proximity detection may be the only viable option, in which case a large number of low-power transmitters are helpful.

6.1.4 Far-Field and Near-Field Radio Systems

Very close to the antennae there is a complex set of electric and magnetic fields, the field strengths of which decay very rapidly (within a very few wavelengths) to the point where they can be ignored. This region is known as the "near field" (in contrast to the "far field" of normal radio propagation). The near field has been recently exploited by RFID technologies and for appliance to appliance near-field communications (NFC). Both are very useful proximity positioning techniques. In RFID systems, the RFID tag is used within the near field that is used first to carry energy to power the device, and second, to carry messages (especially identity messages) by being modulated with information. The best way to think of the near field is not in radio propagation terms but as either a magnetic transformer or an electrostatic capacitor both with large air gaps.

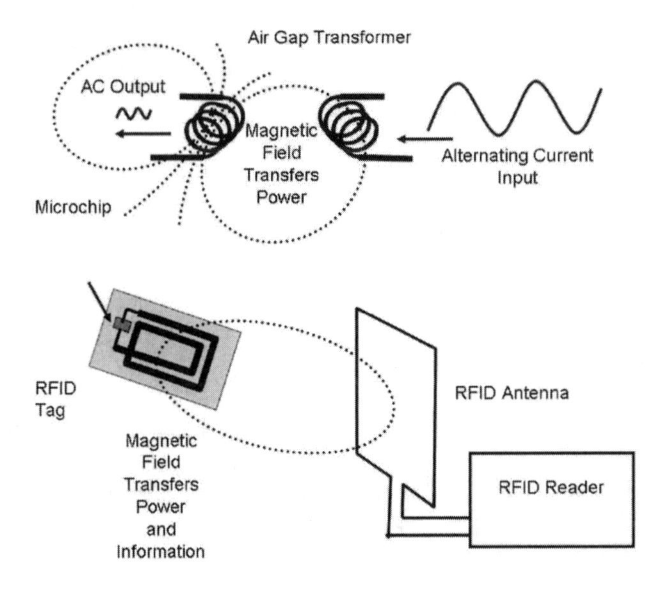

Figure 6.6 Near-field communications by magnetic induction.

If the magnetic approach is adopted, the tag transmitter uses loop (or coil) antennae (as the primary windings of the "transformer") with secondary coils in the passive tags. Figure 6.6 shows how a magnetically coupled RFID tag system (at the bottom of the diagram) works on the same principle as an air gap transformer shown at the top. Both have closed conducting loops that carry alternating electric currents. The magnetic fields induced by the power transmitter are shown as dashed lines and these carry the energy through the air. A classic transformer is made from two adjacent coils but in the case of the RFID tag the coil has been flattened to allow it to be printed (or more likely etched) onto a plastic film. The energy and communications are handled by a custom low-power microcontroller chip.

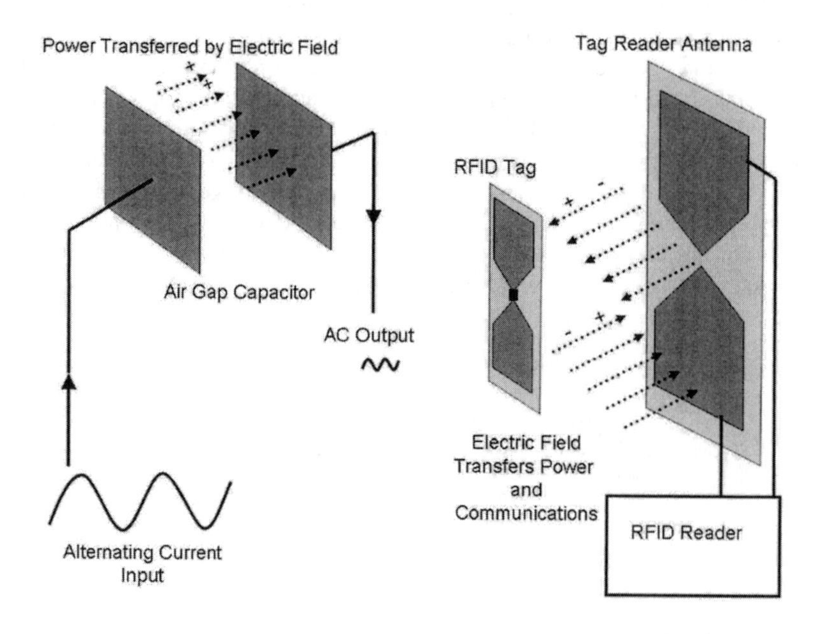

Figure 6.7 Near-field communications using electric fields and capacitors.

If the electric field is used, and this is rare, plates or conducting surfaces are used for both tag transmitter and receiver. The range of magnetic NFC is greater than for electrostatic NFC. Figure 6.7 shows a capacitor with an air gap at the left-hand side of the diagram and an RFID tag system using the electric field on the right. (The tag and tag readers have double capacitors that would be working on opposite phases of the AC signal.)

Backscattering is a longer range approach used by some tags that use the far field where the tag essentially interferes with the radio field, which is detected at the transceiver (note that the term transceiver is used for an integrated transmitter and receiver).

6.1.5 Communications and Sensing

Radio was initially used solely for communications but the reflective nature of metallic objects such as aircraft and ships led to the development of radio detection and ranging or RADAR (explained in the historical Section 1.4.1). The technique can be used to track the position of moving targets by measuring the angle of the signal reflection and the transit time for the pulse of energy transmitted to travel to the target and back again. Doppler RADAR measures relative velocity and uses the Doppler effect (i.e., the change in frequency when objects moving relative to each other send and receive propagating wavelike signals such as the siren pitch change of a police vehicle's siren). Unlike pulse RADAR Doppler (and the somewhat similar radar altimeter) uses continuous carrier radars and employ changes in frequency. There are many other types of RADAR systems beyond the scope of this book, since many techniques are not helpful in the highly cluttered radio environment of urban and indoor areas where the clear lines of sight needed for reflections are rarely available and targets are not metallic. There is one exception, however: that of secondary radar, which does not employ reflections and involves active electronics within a transponder unit attached to the target.

A transponder uses both communications and the directionality of transmitted signals. The simplest transponder is an RFID tag and has some similarities in operational mode to an aircraft transponder. The process starts with a request signal from the base station asking any transponders present to reply (i.e., to transpond). Any unit within range then initiates a transmission back with a modulated signal encoded with identity information. The base receiver can then decode the signal and measure characteristics of the signal. In the case of an RFID tag, it is simply a case of proximity but with aircraft, the secondary system is associated with a pulse system that also gives a heading and range. The transponded code is made to show up on the display next to the blip representing its target.

There are many other transponder systems used in positioning including indoor ultrawideband systems and the tags used for vehicle tolling. It is a very important technique in Whereness and is sometimes present is another form, for example, within the protocols used to set up mobile phone calls. To recap, the main principle is essentially for the base station to ask "Is anything or anybody there?" and if present the person or thing replies with its identity. Smart antennae can be used to gather angle information (angulation) and signal timings to provide distances (lateration).

6.1.6 The Mobile Radio Environment

One of the biggest headaches of mobile radio is the extremely variable signal levels present when radios are moving in a cluttered environment. Although location by proximity is simple in principle, it is inherently unreliable. Figure 6.8 shows the dominant effects. At the top of the diagram the overall effect of declining signal level with distance is shown (as described in equation 6.3). In the center the effect of shadowing due the general topography is shown with effects such as the bending of signals as they diffract around corners. Any obstacle that appears as a "knife edge" vertically or horizontally will cause a sharp diffraction and leads to an unintuitive effect that just because there is a line of sight between transmitter and receiver, one cannot guarantee a good signal. There must also be sufficient clearance at all places along the path due to the interferences within what is known as the "Fresnel zone." [1]

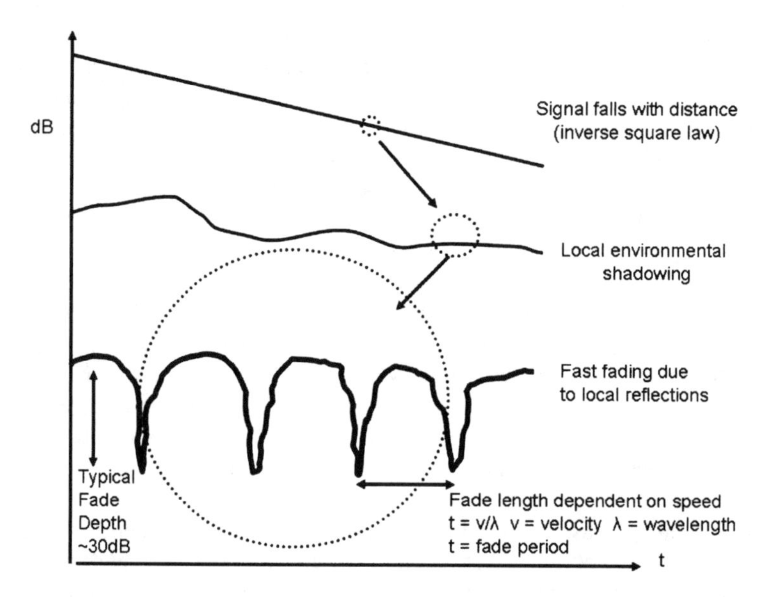

Figure 6.8 Mobile field strength effects.

The signals arriving at a moving antenna may be coming from many directions due to reflections and are known as multipath signals. Standing waves can occur where signal reflections add and subtract so the effect, if moving, is a regular deep fade in signal at the rate at which standing waves are passed. The most dominant reflection is that between the direct signal and a version that has reflected off the ground. These are shown at the bottom of the diagram and can be many tens of decibels deep in level. Fading and all the other variables in the propagation environment makes it very difficult to measure or predict exactly what signal strength will be at any place. The best estimate that can be made will be as a statistic. For example, a mean-level prediction may be made within a square, the sides of which may be many hundreds of wavelengths long but with a variance that is perhaps 30dB. A spot measurement on one day will almost certainly be different when taken at exactly the same place on another day. Another factor moving the levels around is the motion of the antennae especially

[1] The Fresnel zone is a region of space between a transmitter and receiver that takes the form of a solid ellipse. If the edge of any object obstructs the zone, it can cause very significant attenuation of signals. The zone's geometry depends on the frequency of transmissions and the spacing of the end nodes. In the center of the zone, at least 60% of the zone must be clear for a useful transmission.

if they are very directional. Any objects that are moving such as road traffic and objects moving in the wind will also "stir up" the standing waves.

6.1.7 Signatures and Fingerprints

Averaging techniques are the only way to manage the local variability, and a very effective idea is that of collecting radio signal-level signatures or fingerprints. Although a route through an environment will have a widely varying level at any point, on average it will fall within the fade margin. If these margins are collected along a route, a signature for that route can be found, particularly if the route is traversed many times. During a particular journey, it is thus possible to compare the received signal profile with the signature to give an estimate of location.

There are statistical algorithms in the form of filters (i.e., systems that take an input and produce an output that is a function of the input). The Epilogue includes details of the Kalman filter and particle filter that are widely used.

Having looked at general principles we will now consider the positioning potential of the major radio systems available to consumers and businesses.

6.2 Types of Current Radio Systems

6.2.1 System Issues

The section above considered the common basic characteristics of radio relevant to positioning but now we shall consider the specifics of the more important radio systems. The sections will be listed in order of range, starting with satellites in Earth orbit and ending with NFC systems of extreme proximity. There are so many performance trade-offs that it makes little difference in which order to list the systems since no one is inherently any "better" that any other. Range of signal is useful, however, since it brings into focus the very important system issue of networking. In general, the longer range radio systems are more complex since they involve multiple base stations, network control systems, multiple frequencies, and international issues. Indoor low range systems are inherently local and less expensive (but not necessarily less complex, especially in the future).

The most basic radio system is a simple one-to-one link involving a single transmitter and single receiver (e.g., a simple telemetry link such as a remote wireless door bell). The very specific nature of the applications can be useful to Whereness as it is a means of activity and context confirmation.

Broadcasting, or one-to-many links, is useful and includes systems such as aircraft navigational beacons that any number of pilots can use simultaneously. Some interesting new positioning technologies can be piggybacked on broadcast systems [6] such as local digital TV stations.

A more complex broadcasting arrangement is used in GPS and similar systems where a number of separate broadcasts must be picked up simultaneously. To achieve this, a receiver is needed that is effectively a bank of receivers, one per satellite. In reality, only certain parts of the receiver need to be multiplied.

The most complex system arrangements are where there are a host of mobile two-way transceivers, for example, as used in cellular radio. Another class of system of high complexity (but not yet to the degree used in cellular radio) is the hotspot Internet access system (such as WiFi). One of the most common aspects of complex communication systems involves "trunking" (i.e., the tree and branch-like approach that is used to configure communications networks). Trunking can be very useful to Whereness since part of the task of a computer system managing trunking is to keep account of which end terminal is using which common resource that usually has a physical location.

6.2.2 Global Navigational Satellite Systems (GNSS)

GNSS is a generic term that covers systems like GPS, GLONASS, Galileo, and their predecessors and successors. The phenomenal uptake of GPS for business and consumer use has brought the former military technology into the public's mind so that it is likely that the overall future of GNSS will be for both military and nonmilitary applications. This is a situation not unlike the evolution of two-way digital radios which has now evolved to the stage where the nonmilitary systems are ahead in capabilities.

6.2.3 GPS Infrastructure

We will start with an overview of GPS but it is worth noting that it is not so much about how GPS works but more about what can be expected by users from the system. If a more in-depth explanation is needed, there are many excellent publications, of which Kaplan [7] is particularly comprehensive.

GPS is a complex system and the receiver module, embedded within a domestic satellite navigation (sat-nav) unit or cell phone, is only one small part. Overall the system consists of a constellation of 24 operational satellites (plus spares and any failed units) in mid-Earth circular orbit, controlled by a network of ground stations using two-way radio links. The height is such that each satellite orbits once every 12 hours and the configuration of the satellites is such that they are arranged to be well spaced in the sky so that from the ground one can expect to "see" up to an average of 12 satellites above the horizon at any time.

A receiver needs to be able to receive signals from at least three satellites to find a two-dimensional position (or four for a position that includes height). If any local buildings, tree foliage, or anything else opaque to the microwave radio signals is present and occludes a clear line of sight, signal reception is degraded limiting the potential for a position to be computed. This is a frequent problem in dense urban areas where "urban canyons" prevent enough of the sky being available. Since the satellites always appear to be slowly moving across the sky, a location may work at one time but not at another. The signals are quite weak (when compared with other domestic radio systems), so good quality antennae management is needed for reliable use.

People take the availability of the service for granted but it has a number of potential problems. First, it was built and is paid for by the U.S. Department of Defense and is under the political management of the U.S. government. The initial decision to open GPS in the early 1980s for nonmilitary use was taken after an international Cold War incident when an airliner in the wrong place was shot down. It was felt that it was wrong to deprive basic GPS when perhaps people's lives could be saved by better civilian navigation systems. After the end of the Cold War in 2000, the system was improved for public use by the removal of an artificial degradation (called selective availability, or SA), which improved performance by an order of magnitude. At any time, however, the system could theoretically be either degraded or denied, especially in times of crisis or in regions with conflicts.

Second, aircraft may not rely on a single GNSS because of the inherent unreliability of a single complex navigational system (three would be needed so that if one failed the other two could check each other's outputs). Such a situation could occur if there was a catastrophic fault, for example, with a number of satellites due to natural cosmic disruptions, or perhaps acts of space warfare, or with the ground station's failure (without which the satellite's accuracy would soon drift).

Finally, GNSS are radio systems are prone to all the problems inherent with all radio systems. Perhaps the biggest threat is that of service denial by interference that may be deliberate. GPS and similar systems use wideband spread-spectrum broadcasts and are more difficult to jam than conventional narrowband domestic radio systems, but radio jamming may become a problem similar to computer hacking, especially as jamming equipment becomes more sophisticated. To sum up, by all means use and benefit from GPS but keep mindful of what is behind the scenes, politically, economically, and technologically, because each area has many potential problems that could disrupt GPS in the future. Assuming, therefore, that a user has access to four viable satellites, how does the position get fixed?

Figure 6.9 Miniature GPS module (left) with patch antenna (right).

6.2.4 GPS Positioning

The satellites broadcast a number of signals on the two (current) frequencies in the microwave band at 1575.42MHz (L1) and 1227.60MHz (L2). The former is used by the vast majority of public users and at this frequency, small patch and rod antennae are possible, which make portable equipment convenient. The timing of each satellite's transmissions is governed by an on-board cesium beam atomic clock, the performance of which is maintained by the ground stations that monitor system quality. Code Division Multiple Access (CDMA) modulation is used, which is a form of digital communications. It employs a set of unique pseudorandom sequences of binary pulses to differentiate between the satellites (which all transmit on the same radio frequency channel). In the receiver, after amplification, although all the available signals are present together, a correlation process selects the unique codes by comparing them with known locally generated replicas. This process results in a coding gain that effectively amplifies the level of the wanted signal and rejects the unwanted. The timing of the pulses can then be used to calculate satellite distances.

Each satellite visible therefore has a locus of equal distance that is spherical and where a number of these spheres intersect, the receiver location can be calculated. Systems such as GPS that use signal time differences are known as circular lateration systems (and can be contrasted with the systems using hyperbolic lateration based on signal phase differences).

Knowing the satellite distances is important but to find local position it is necessary to know (with respect to the center of the Earth) the precise position of each satellite. The ground stations measure and calculate this information that takes the form of a set of parameters (the GPS ephemeris), which contain the orbital elements (or Keplerians). In addition to the timing codes, the satellites also relay the ephemeris data to users on a low bit rate data channel also modulated onto the carrier as the broadcast message. Before the positioning algorithms can begin, an up-to-date ephemeris is needed. It can take several minutes for all relevant data to be received so the initial session can be lengthy (which is known as a "cold start"). Receivers store this data in nonvolatile memory so that if a new session is started again before that orbital information has significantly drifted, the startup time can be very rapid (perhaps a few seconds).

It is also possible to relay the ephemeris data to the GPS receiver via a non-GPS route, for example, by a cellular radio data signal. This arrangement is part of a methodology known as assisted GPS (AGPS) and can accelerate greatly the time taken for a cold start and is an essential part of the E911 emergency calling system.

Matrices are used to solve sets of simultaneous equations containing the timing information received from the available satellites and after processing, the positional information can be estimated. It is expressed in various formats (e.g., latitude and longitude in degrees, minutes, seconds [or using decimals], height in meters above mean sea level, and velocity). The geodetic standard WGS-84 is used as the model for the approximate shape of the Earth. Many GPS receiver modules have a navigational output signal (via an RS232 port), and will present their data according to the standards defined by the National Maritime Electronics Association (NMEA). This is a simple text-based protocol that generates standard "sentences" containing fields of latitude, longitude, time, altitude, velocity, satellite availability, and quality information. Proprietary formats are also available. These modules greatly facilitate the embedding of GPS into consumer appliances very cheaply.

The accuracy of the readings varies according to a large number of factors, including system limitations such as clock quality and code parameters, atmospheric propagation anomalies, and everything that affects local signal strength in the local environment. A typical domestic receiver should be giving a positional performance better than 3m most of the time. By increasing the sensitivity of the receivers (with more sophisticated correlation techniques) it is now possible to use the GPS devices, at least to some extent, indoors, especially near windows. It is remarkable that GPS transmitters in orbit, more than 20,000-km distant and transmitting only 27W of power (about half the power of a conventional incandescent electric lamp) can still be providing useful positional information indoors. Figure 6.9 shows how small a typical GPS receiver module has now become.

6.2.5 Timing, Differential GPS (DGPS), GNSS Augmentation Services, and Surveying

There are many specialist applications for GPS. The accurate local timing that is effectively locked to a network of atomic clocks can be used as a substitute atomic clock for other radio applications. Some of these, for example, may be positioning systems based on terrestrial radio.

Differential GPS (DGPS) is a technique whereby the overall performance locally is improved, typically, by an order of magnitude. A local base station receives GPS signals and calculates offsets for each satellite with respect to real position that has been surveyed (this is known, a priori). This set of offsets, or differentials, is then communicated to local GPS receivers that often have an extra input for the DGPS service. DGPS can be part of an AGPS service or offered to specialist users, for example, fishing fleets.

A similar approach is being offered to improve the overall quality of service by augmentation services. Examples include WAAS in the United States operated by the Federal Aviation Administration (FAA) and EGNOS in Europe operated by the European Space Agency (ESA). Networks of ground stations monitor the GPS signals and calculate corrections for atmospheric disturbances, timing errors, and other system disturbances and record these so that historic readings can be improved. EGNOS is the first part of Galileo to be operational.

Land surveying in the construction industry, by mapmakers and geoscientists, can also use GPS but in a slightly different way. Measurements based on the phases of GPS carriers known as carrier phase enhancement GPS (or CPGPS) can provide centimeter accuracy (enough to measure continental drift). This can be achieved in either real time through delivering correction messages to the user's receiver via a radio, Internet, and so forth, or in software during postprocessing. Regional networks are being established that provide dual frequency phase corrections to user communities, for example, the OS Net™ network in Great Britain. Although much better than the 0.3m accuracy expected from the military GPS signals, the trade-off is that it is quite slow and not suitable for guiding weapons (the main reason why the military encrypts their higher specification GPS channel).

6.2.6 The Future of GNSS

It is looking likely that by around 2015 there will be at least three fully operational GNSSs operated by regional powers. GPS III is the planned evolution of GPS with extra channels for civilian use and enhanced performance (i.e., bringing GPS up to a similar performance to the E.U.'s Galileo proposals).

The Russian government has committed to upgrading GLONASS until it has a full working constellation, and discussions on compatibility are ongoing. At the fall of the Soviet Union the system degraded until coverage was virtually unusable. The Indian government is cooperating with the upgrade.

Many other nations are participating in the Galileo system, which (at the time of writing in December 2007) is looking very likely (but not guaranteed). The original plan for a public-private partnership GNSS failed, and now public funds will be used to develop the system, ready to be exploited by commercial industry.

Galileo is not a military system and has defined services useful to commerce, public infrastructures, and the general public. There is a basic unencrypted free service and then several others aimed at specific user groups. The main difference is that these services have quality information included so as well as knowing where something is, it will also be possible to know the accuracy and integrity of the broadcast information, providing a legal guarantee of the system's quality. This is an important benefit if monetary transactions or safety critical activities are dependent on the raw information. The following is the proposed list of Galileo services as listed by the European Space Agency (ESA) [8].

- Open access (free to air);
- Commercial, high performance with fees (encrypted);
- Safety of life (guaranteed accuracy);
- Public regulated service (for government agencies, encrypted);
- Search and rescue (includes two-way communications system for beacons).

Although one might consider the idea that the various GNSS systems would be competing, in fact it could lead to a situation where the benefits are greater than the sum of the parts. Provided reasonable receiver compatibility is possible (i.e., many parts of a multisystem receiver would be common or shared), overall a user could expect vastly better coverage and availability for the much greater numbers of satellite measurements that can be used in calculations:

- GPS, 24 satellites;
- Galileo, 30 satellites;
- GLONASS, 24 satellites.

Clearly, the future of GNSS is dependent on launch technology and its costs. All quasi-governmental space projects are expensive but involve a degree of politics including maintaining strategic technological capabilities and support for the aerospace and defense sectors. In spite of a general move towards the civilianization of GNSS, other agendas are present.

Perhaps the wild card may be the emerging business of space tourism and the concept of cheap reusable spacecraft. It may be possible to use these craft as launch platforms for low Earth orbit microsatellites. Microsatellites [9] have been developed and involve short-lived, low specification, small "disposable" satellites that may be launched in high numbers at very modest costs. It is feasible that an "Internet in the sky" may be deployed using microsatellites within a decade, and it is likely that these will also serve as positioning, imaging, and environmental sensing devices. Signal strengths could be large and in building coverage more feasible than with conventional GNSS.

6.2.7 Dedicated Terrestrial Systems

There are many application-specific and dedicated (often proprietary) radio positioning systems, some of which use techniques developed during the Second World War (see Section 1.4.1). Long-range navigation, or LORAN, is still in use (in the form of the LORAN C standard) and uses VHF radio beacons that transmit their carriers over coastal regions. LORAN (and the now obsolete Decca system) uses carrier phase to determine position. The locus of equal phases between two transmitters is a set of hyperbolae. If several pairs of transmitters are in range, the intersection of the hyperbolic loci gives position. Accuracy is worse than GPS by about an order of magnitude (i.e., 10s to 100s of meters) but adequate for many marine and some land uses. LORAN C is still operational and is being upgraded. Although prone to radio atmospherics it is a simple system and its robust nature makes it complementary to GNSS.

In a similar vein are the various aircraft systems such as ADF and VOR. Automatic direction finder (ADF) allows a pilot to know the bearings of dedicated VHF navigational beacons so that from several bearings a position can be plotted on a map. A simple loop antenna can be tuned to find the direction of each signal to which the receiver is set.

VOR (VHF omnidirectional range) is more complex. The beacon transmits an omnidirectional reference signal and a second transmission from a rotating highly directional antenna, the phase of which is electronically varied according to the absolute direction. The receiver can calculate its heading from the phase difference it is experiencing, which will change as the flight continues unless heading directly in line with the beacon.

Another dedicated class of system is that used to track or follow targets that may be animals, fellow radio amateurs, orienteering enthusiasts, or perhaps stolen vehicles. In these cases the transmitter is small, simple, and narrowband (to preserve battery power). A simple pulsed "bleep" will conserve energy further. When triggered, a moving receiver is used to find the direction (i.e., direction finding, or "DF-ing"). Although handheld directional antennae are used, another approach is to use a simple static phased array mounted on the roof of a vehicle. These can be seen on police vehicles as a square pattern of four quarter-wave monopoles. These are electronically switched in phase, which can simulate the effect of rotating the array. The phase switching is synchronized to a display to show the approximate heading of any received signal. A similar arrangement is used for coastal maritime radio communication (see Figure 6.10).

The latest manifestations of the DF technique are emerging for use with the ISM band (the band used by WiFi at 2.45GHz). Small tags can be attached to objects [10] that can be easily lost about a room (e.g., vehicle keys). When the small remote controller is activated it "pings" the tag, which is a transponder that replies. The reply is picked up by the remote, which displays the approximate direction of the signal and its level.

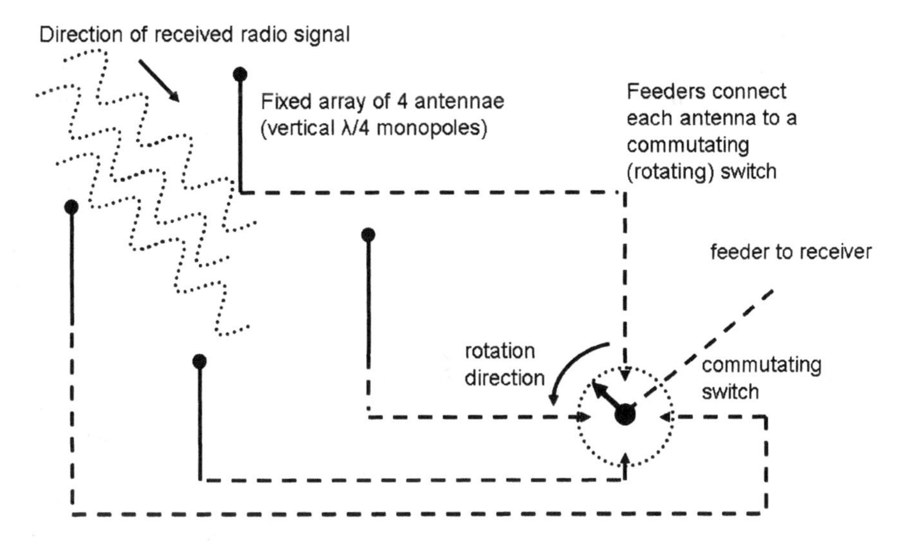

Figure 6.10 Simple phased array to determine angle of arrival of radio signal.

6.2.8 Trunked Radio and Cellular Radio

When cellular radio evolved from fixed telephony, the trunking (which was performed in a hierarchy of exchanges or switches) had to be modified to deal with the problems of radio. Radio frequency bands (which are inherently scarce resources that are highly regulated, allocated, and more recently auctioned commercially) must be tightly controlled and managed with maximal spectral efficiency. In other words, operators and their systems seek to maximize the number of uses per square kilometer for any given block of spectrum (without disrupting by interference any other users).

The way this is achieved is to allocate radio channels dynamically under computer control to those who need them on demand from a common pool. The channels can be reused but only at distances where the energy from transmissions has dropped to a level where it is negligible (or "in the noise"). This type of interference is known as cochannel interference and can be engineered by carefully sighting base stations, antennae, and controlling power levels. The term "cellular" became common as the (theoretical) pattern of channel reuse can resemble the cells of a honeycomb.

When a user makes a call a local base station is allocated a free channel that will have a frequency and may have a time slot or code depending on the standard. As the mobile transceiver moves the call or session is handed over to an adjacent cell and a new channel is allocated seamlessly. When no calls are being made, the central system still keeps a record of the local users by making regular silent calls to check on the status. This is necessary so that outgoing calls can be routed efficiently. If there were no location registers all users would need to be contacted from every base station, everywhere, which would waste valuable resources. The location register is therefore consulted by the system before contact is attempted to find the nearest base stations to the user.

There are two parts to location registration. First, a home location register keeps account of all details of each user. Second, a visitor location register keeps account of those mobiles visiting an area. Details are swapped between registers as mobiles roam. We can see, therefore, that a coarse Whereness system is in operation keeping the positions for every active unit at all times.

Some location-based services (LBS) are based on location register information (or Cell ID) and there is potential for many more. As cells get smaller, in particular in dense urban areas, the accuracy improves but accuracy is poor in very rural areas where the cells may be tens of kilometers apart.

There is, however, a complementary situation between cellular positioning and GPS. In dense urban areas GPS is likely to be poor in performance due to high-rise buildings, but cellular coverage is very good and cells closely spaced. In contrast, GPS works very well in open country where cell base stations are sparse.

Although Cell ID is crude it comes at no extra cost other than the requirements to make the embedded information available to interested parties. With some systems (e.g., GSM) it is possible to use other, more accurate system information. GSM is a time division multiplexed system where each physical radio channel is shared between up to eight users who are allocated timeslots on the carrier. To ensure that information from each mobile user is transmitted at exactly the correct time, a "timing advance," or TA, control system is used. When a mobile first contacts a base station, it has no information concerning its range and the time its signals will take (at light speed) to reach the base station. It therefore only transmits an access burst in the center of a time slot and leaves a substantial timing guard bands unmodulated. The base station detects approximate range of the mobile from the time offset between the incoming access burst and its own timing multiplexing. It then sends a correction factor to the mobile so that all further time slots are aligned and no resources wasted on unnecessary guard bands. As a call progresses the timing advance is constantly updated and corrected.

It can be seen, therefore, that approximate range information is available and some attempts have been made to use this information for positioning. For example, if the network forces a mobile to register with three different and geographically separate base stations then triangulation can be used to fix position. Further improvements may be made by the enhanced cell global identity

(ECGI) technique, which includes signal strength measurements of the mobile's signals and the enhanced observed time difference and uplink time difference of arrival (EOTD and U-TDOA) techniques, which compares the arrival time of signals from base stations to mobiles. In U-TDOA, the positioning is centralized within the network whereas EOTD uses localized autonomous positioning within the mobile unit. The trade-off of these extra positioning enhancements is the complexity and investments required. Overall a typical system would at best be giving positional accuracies of around 50m [11, 12].

Angle of arrival (AoA) measurements can be taken by using highly directional antennae at base stations using phased array beam-forming techniques (developed from military and satellite systems). In the first cellular systems, simple sector antennae were used, for example, to illuminate three sectors, each with 120°. The bases were placed not in the center of cells but at the corners of a nominal honeycomb-shaped coverage pattern. These arrangements were fixed, but with narrow spot beams that can be made to move (by automatic signal phasing) and track targets, AoA information becomes more useful.

The biggest problem with all the cellular techniques for Whereness is that they are not standardized. Some standards have emerged but they are optional. Although an operator may offer an excellent and useful location-based service (LBS), it would be unavailable to users who roam onto competitive networks and possibly for visitors, for whom the equipment may lack vital software or configuration. An example would be a network of location monitoring units (LMUs) providing the means to measure signal timings at dedicated locations. Access to the measurements would clearly be restricted to those who are customers of the operator who has made the investment. Given the fragmented approach to cellular standards in many global regions, it seems unlikely that a unified positioning standard will emerge.

The most useful approach would be to treat all and any system positioning services as inputs to a more general Whereness service with convergence functions for each network (see Figure 3.2).

6.2.9 Hotspots and WiFi Positioning

Hotspots are the colloquial name for Internet wireless access points that are now very common in most built environments and provide static broadband connections with typical ranges up to 100m. The positioning techniques discussed in this chapter are all applicable to this relatively new form of communications and some interesting new approaches are emerging, but the most useful opportunity is the ability to gain an accurate position indoors.

Most current hotspots conform to the WiFi standard (802.11g/b) operating on the unlicensed industrial, scientific, and medical (ISM) band. The technology is developing rapidly and various enhancements are expected including the opening of other bands, some of which will be licensed to specific operators. A major development is the increase in power leading to systems such as WiMax, which operate at extended ranges (typically 1km).

Generally it is assumed that access base stations are fixed in position and the user transceivers are portable but used while not in motion. WiFi and similar protocols used by hotspots generally do not have the extra data protocols needed to eliminate errors when truly mobile (in contrast with true cellular radio, which contains significant data redundancy and protocol overheads).

Simple proximity is the most basic approach to positioning (similar to the Cell ID approach) and a number of products are available that use this technique [13, 14]. More advanced systems use fingerprinting and statistics. Measurements are taken of the signals from all hotspots in range (and typically may be 10 or more) and from past records the likelihood of position is calculated [15]. Intel's Place Lab pioneered the technology of WiFi positioning and a suite of software is available for experimental use that uses advanced techniques (e.g., the particle filter). In order to collect a large number of mapped hotspots, it reuses the wardriving records that have been collected by amateurs (see Section 2.3.7).

Lateration based on timing is also possible but is not widely available. Signal angles can be measured and this approach is gaining momentum since the advent of MIMO technology. Multiple input multiple output (MIMO) devices use advanced antennae to improve communications performance.

The main drawback of the statistical approach of measuring signals a priori is the effort it takes both to collect the initial survey and to update it should the local environment change. For example, if an open plan office was mapped and then a row of metal cupboards moved, the disturbance in the signal strength field would be significant. It would be ideal to have an indoor system that worked rather like GPS, using highly accurate time of flight pulse timing. Advances in semiconductor devices have made this approach a reality quite recently in the form of ultrawideband systems (UWB).

6.2.10 Ultrawideband Positioning

Until this technology was developed the only other approach to accurate indoor positioning used ultrasonics (details of which are in Chapter 7). The term UWB is used mainly to describe a new class of low-range radio communications where the bandwidth is very high (typically 500MHz or 20% of the carrier frequency) [16]. The first UWB systems are the ones of interest to positioning since they use nanosecond pulses of radio energy, which are ideal for accurate timing applications such as positioning and imaging. More recently the term has included systems with more conventional modulation that are not especially useful for positioning.

The idea behind nanopulse UWB is radical and disruptive and is causing the radio regulations to be modified to take into account its special characteristics. In essence, it is a very simple idea that concerns the replacement of the normally complex modulation and demodulation system of a digital wireless with a simple pulse generator (similar to the idea behind radar). A pulse of energy very narrow in time has a radio spectrum that is very wide. At the transmitter, the energy can be thought of as being transformed from a time domain nanosecond wide pulse signal into a very, very low-level frequency domain signal that spreads across a significant part of the radio spectrum. If care is taken, the level is lower than the ambient noise (and confusingly below the level to actually be regarded as interference). At a UWB receiver the received spectral components are "despread" using correlation techniques (along the same lines as for a GPS receiver) and the pulse is recovered. Information is encoded by the presence or absence of pulses with respect to time but the actual epoch of the pulses, if synchronized and transmitted from multiple base stations, can operate in a similar way to a GPS system. UWB tags can thus be positioned in the overlap area between a set of transmitters or receivers and positioning performances of centimeter accuracy in three dimensions is expected.

Currently commercial units are emerging [17] and in time will probably be combined with more conventional hotspot base stations. Applications are being used in environments such as hospitals, where there are scarce high-value assets and specialist workers, where good logistics can be lifesaving.

6.2.11 Low-Range Radio Systems (Bluetooth and ZigBee)

A number of useful radio standards have been developed for applications where range is less than the 100m normal with WiFi. These systems can also be used for positioning using the same approaches as for other hotspots mentioned above. For applications such as remote earphones and other communications about the person, the cell phone manufacturers pioneered the development of Bluetooth. For lowrange telemetry (i.e., machine-to-machine communications), a different protocol called ZigBee has emerged. All can share the same unlicensed spectrum.

A future building could thus have several useful infrastructures that could support Whereness applications. WiFi would be available for general wireless broadband Internet access ZigBee for the control of the building lighting, heating, fire and intruder alarms, and energy management; and islands of Bluetooth associated with personal equipment. The very low ranges (3-10m is typical) result in quite accurate positioning by simple proximity.

6.2.12 Dedicated Short-Range Communications (DSRC) Systems and Active RFID

DSCR systems are used at the roadside for vehicle applications such as tolling, parking payments, and between vehicles for more advanced concepts such as adaptive cruise control. The radio band allocated is in the microwave region (5.8 GHz) and as such uses conveniently small antennae suitable for easy integration into physical tags. The technology is part of what can be considered to be a radio frequency identification (RFID) system, since a principal function of a tag is usually to identify itself for validation. The activity of tag reading creates a time and position stamp which is then used in various enterprise software applications, for example, to collect bills. In systems where every vehicle is tagged, the DSRC can provide traffic flow information and also play an important part in crime prevention and detection.

DSRC tags are transponders (similar in operation and principle to military IFF systems discussed in Section 1.4.1) that handle a short two-way data "conversation" and are therefore considered to be active tags. The range can be in the order of tens of meters and therefore a power supply is needed. This is in contrast to passive tags that are powered by the near-field communications radio signals. Many cities globally have installed tolling systems (e.g., Singapore and Toronto) and they are one of the most common ITS applications and a very useful part of Whereness that could be exploited further.

6.2.13 Passive RFID

RFID technology is a rapidly growing business and a truly disruptive technology that is changing the methods used in many logistics, access security, and fee collection systems. As with the longer-range active tags discussed above, the activity of tag reading provides an unambiguous time and position stamp. Passive RFID tags are not passive when they are being used since they pick up power from the local radio frequency field (usually using the near field), which energizes the electronics in the microchip attached to their antennae. The anatomy of most tags is similar. A loop antenna encompasses most of the tag (which is usually a few cm square), and there is an associated capacitor that forms a tuned circuit resonant at the tag frequency. There are many RFID frequencies allocated worldwide but the only common dedicated band is the high frequency (HF) band at 13.56MHz.

A typical large-scale system is the Oyster Card [18] used on the London Underground rail network, where thousands of tag readers are installed at ticket barriers and are used by millions of passengers daily. The cards are contactless but must be placed in close proximity to the tag reader mounted on the barriers. The tags can be reused since the tag can be recharged with cash via an Internet service.

Tag readers are actually transceivers and the tags contain a small amount of data storage (1kb is typical) so that they can store information about available credit and personal details. Normally, RFID tags contain pointers to data located on databases elsewhere.

Whereas fee collection and building access systems use fixed tag readers and mobile tags, it is also possible to use an RFID system the other way around. For example, a mobile container crane may need to be accurately positioned along side a dock or rail head or roadway in order to safely transfer shipping containers between ships, trucks, and wagons. In this situation a mobile tag reader could be fitted to the underside of the crane and the tags embedded under the concrete roadway at very specific positions. When the crane moves, the tag reader would be used to indicate exact alignment that can be made to within a few centimeters. Tags used in wet situations, which are encountered in both industrial settings and also when tags are embedded within living tissue (for example, with animal tags), require the use of a much lower frequency band. Low-frequency (LF) tags operate at 150kHz, and significantly above this frequency passage through water attenuates signals and makes them unusable. The main problem of very low-frequency tags is the need for an antenna that is wound around a ferrite core. This is an expensive and bulky component but is necessary to receive enough power at the very low frequencies used.

The application of RFID is often limited by the physical constraints of the radio environment. Metal is generally opaque to radio signals and can also interfere with field patterns. Tag antennae are often bulky and need to be carefully positioned. Interference, in common with all other radio systems, is another concern especially if signals are being deliberately intercepted for criminal purposes or jamming is used to deny service.

Another limitation on tagging technology in general is the issue of personal privacy. People are increasingly reluctant to trust organizations with personal information and the automatic nature of tag reading does nothing to help people have confidence that the information is being used in their best interests. It is likely that this problem will increase unless the tag industry and corporate operators of systems inject enough transparency into operations to give the public confidence. A more general Whereness service, where users can see exactly what is known about their activities and by whom, might lead to a reduction in their concerns. An even better situation may result if they are also allowed to edit their own tag profiles.

6.3 Summary

This chapter was about radio-based positioning and started with a brief introduction to radio theory to help explain how the main three ways to find position operate and what radio parameters are important. First, radio can be used to find distance (lateration) by timing signals. Pulse flight time is the most common method and it is used in GPS, radar, and many other systems. The other approach compares the phase of radio frequency carriers. Timing signals between three links allow position to be found at the intersection of three imaginary spheres centered on the transmitter. With phase measurements, points of equal phase are hyperbolae hence the term hyperbolic navigation, which is used to describe systems such as LORAN.

The second technique (angulation) uses directional antennae to find the angles between the transmitters and receivers and is widely used in radar and radio imaging. The third approach, and the one used in many current cellular radios, WiFi hotspot and ubiquitous computing applications, is using simple proximity.

Either the radio is there or not; if present, signal strength may give a very rough approximation of range. The radio environment propagation statistics are very complex, especially within the built environment, so finding position on signal strength alone tends to be unreliable.

The importance of bandwidth and signal-to-noise ratio was discussed. More power is needed for a greater bandwidth signal (i.e., a signal modulated with more information). And when accuracy is needed, the greater the signal-to-noise ratio, the better, since noise (or random signal-level fluctuations) gets translated into timing errors.

Near-field communications (NFC) was discussed since this mode of radio is now widely used in RFID systems and has the advantage of not only providing communications and positioning, but also transfering power to operate a simple RFID tag's electronics.

A review was made of all the most common radio systems including GNSS, cellular radio, hotspots, and radio tags. GNSS is set for a major enhancement in the medium term when GPS is joined by the EU Galileo system and others. Accuracy, availability, and resilience will be increased and commercial services will be launched offering service guarantees.

The most recent transponder tags use ultrawideband nanopulse communications that is perhaps the most interesting new radio technology for positioning indoors. It has very high accuracy combined with all the other advantages of radio.

As the wireless becomes more widespread as the scope of the digital networked economy increases, truly ubiquitous positioning looks increasingly achievable in the medium term.

References

[1] Straw, R., Dean, Ford, S.R., and C.L. Hutchinson, *The ARRL Handbook for Radio Communications 2006*, American Radio Relay League (ARRL).

[2] Amos, S.W., *Scroggie's Foundations of Wireless and Electronics*, 11th Edition, Newnes, Oxford, UK: Boston, MA: 1997

[3] Ralph, D., Searby, S., (eds.), "Location and Personalisation," *BT Technology Journal*, Volume 21, No. 1, Jan. 2003.

[4] Bross, M., Smyth, P. (eds.), "Telecomms Unplugged," *BT Technology Journal*, Volume 21, No. 3, Jul. 2003.

[5] Dennis, R., Wisely, D., "Mobility and Convergence," *BT Technology Journal*, Volume 25, No. 2, Apr. 2007.

[6] *Rosum*, http://www.rosum.com/, Jan. 2008.

[7] Kaplan, E.D., *Principles and Applications Understanding GPS*, Norwood, MA: Artech House, 1996.

[8] *ESA The Future—Galileo*, http://www.esa.int/esaNA/galileo.html, Jan. 2008.

[9] *Surrey Satellite Technology Ltd.*, http://www.sstl.co.uk/, Jan. 2008.

[10] *Loc8tor Ltd.*, http://www.loc8tor.com/, Jan. 2008.

[11] Drane, C.R., Rizos, C., *Positioning Systems in Intelligent Transportation Systems*, Chapter 7, "Non-GPS Systems," pp. 237-271, Norwood, MA: Artech House, 1998.

[12] Salmon, P.H., "Locating Calls to the Emergency Services," *BT Technology Journal*, Volume 21, No. 1, Jan. 2003, pp. 28-33.

[13] *Vocera*, http://vocera.com/, Jan. 2008.

[14] *Ekahau*, http://www.ekahau.com/, Jan. 2008.

[15] *Place Lab*, http://www.placelab.org/, Jan. 2008.

[16] Gu, X., Taylor, L., "Ultra-Wideband and Its Capabilities," *BT Technology Journal*, Volume 21, No. 3, Jul. 2003, pp. 56-66.

[17] *Ubisense*, http://www.ubisense.net/, Jan. 2008.

[18] *Transport for London*, Oyster online, https://oyster.tfl.gov.uk/oyster/entry.do, Jan. 2008.

Chapter 7

Sensing Position Without Radio

7.1 Alternatives to Radio Positioning

This chapter discusses all the nonradio methods to find position. Chapter 6 considered radio positioning, which has been the historical method and has many advantages, especially the ability of radio signals to pass through many materials. But radio also has some serious disadvantages, which is why it is important to consider alternative strategies. Sensing is a common thread in this chapter, and it is important to note that sensors and sensing are a rapidly growing area of both science and technology. Humans and animals have evolved a keen sense of position but radio is (as far as we know) never used. All natural positioning is based on sensing, by which we mean the collection of information from sensors that convert physical phenomena into information that can be processed.

The main artificial methods align well with nature and include optical sensing, sound sensing, movement and pressure sensing, and magnetic and electrical sensing. Nature also uses techniques that technology has yet to master, for example, using chemical sensing and olfactory sensing and the area of biomimetics (i.e., copying nature) is proving a rich resource for commercial exploitation.

The great strengths of radio also are associated with its problems, in particular, the problem of interference. Strict radio regulations limit emissions (and reception is limited in some instances but mostly from a political perspective), but these restrictions do not generally apply to the other methods of positioning. There are health and safety issues but there are no communication licenses required for the use of, for example, infrared or ultrasonic systems. Mechanical systems such as odometers, pedometers, accelerometers, and gyroscopes are obviously not licensed. Many of these technologies have versions that are also very cheap and simple. Radio has always been a difficult technology and the scarce radio spectrum has led to exacting developments. In contrast, many sensors are single devices connected to the simplest of processors.

The research area of ubiquitous computing was introduced in Section 1.4.6, and it is important to remember that most networked artifacts of the future (which some would say would include almost everything) will include three aspects: computing, wireless communications, and sensing. Recently, a number of new consumer devices such as the Apple iPhone and the Nintendo Wii game machines have drawn attention by their ability to sense the position of the user's hand holding the unit. Robotics is also an advancing area of technology that is pioneering the use of position sensing (see Section 1.4.13).

The next chapter will discuss the importance of maps but it should be remembered that until very recently when satellite positioning became available, cartographers used sensing techniques to map the world. Optical observations of horizontal and vertical angles with theodolites and distances with tapes, chains, or distance wheels were taken laboriously in the field.

Sensing movement using mechanical methods and using inertial navigation do not rely on any external phenomena and are wholly autonomous, which is why the techniques were developed for military purposes where external signals are a source of vulnerability. Sophistication can vary enormously from the simple vehicle odometer or pedestrian pedometer to the optical gyroscopes used in aerospace and missile systems.

A final class of techniques not involving radio concerns systems associated with ICT and telephony, where the physical connection to a wired network or its use is an indication of location. Also, the use of information that is, a priori, entered in personal calendars and diaries that can then be confirmed, a posteriori, by network use. The advantage of using basic ICT components for Whereness is that very few extra investments are needed in infrastructure since most of the intelligence is based on software.

7.2 Infrared Systems

Infrared (IR) is optical but uses light waves that are longer in wavelength than the visible spectrum (visible in a rainbow). The longest visible light is at the red end of the spectrum at around 700nm and the shortest radio wavelength is about 1mm. Infrared is the portion of the spectrum in between. Most radio signals are created coherently, which means that the signals at any frequency have continuous phase. The optical equivalent is lasers but optical receivers are incoherent. They involve sensing the presence of energy regardless of its format, whereas most radio receivers involve a coherent demodulation process. This is the reason why optical systems are cheap and easy but the downside is that spectrally, they are inefficient. For example, if several IR remote controllers are used at the same time in a room to control TV and HiFi, although the codes transmitted will prevent false programming, they will tend to block each other's signals. Only one IR transmitter can be used unambiguously. Optical fiber communications gets around this problem by a wavelength division multiplexing (WDM) where each signal is effectively a different "color" of IR but signals are still received incoherently. However, by careful pointing and coding it is rarely a problem. True coherent optical radio is unlikely so IR will remain a cheap, crude, but very useful positioning tool.

IR transmitters are light-emitting diodes (LEDs) where the two layers of semiconductors used in the device are chosen to emit IR light when a small electrical current is passed. They have relatively high bandwidths and can thus be pulsed on and off very quickly (in the MHz region). IR optics can be used to focus narrow beams, and large parallel arrays of diodes can generate bright (but invisible) illumination of scenes. This is a common feature on smaller surveillance cameras that are sensitive to near IR as well as visible light.

There are three approaches to detecting IR. The first method is using a simple IR detector diode that is made to pass more current or generate a voltage when illuminated by an IR beam. The second method is to use a camera sensitive to IR where a large array of semiconductor detectors, fabricated as an integrated circuit, has an image focused by a lens. The third approach is passive, using a passive IR detector (PIR) where far infrared (i.e., heat radiation) is detected by special material whose surface reacts to changes in intensity. Intruder alarms use this technology to detect warm bodies and some sophisticated military cameras also integrate the materials with more conventional silicon chips. In all cases the presence or absence of IR radiation can be used to sense the presence or absence of objects, people, and animals in a specific place. Positioning is thus by proximity with accuracy being determined by the physical geometry of the system. Overall ranges are low (tens of meters) and are used mostly indoors.

7.2.1 Indoor Positioning

A citywide outdoor ITS experiment that used IR beacons was described in Sections 2.2.1 and 4.2. The beacons were placed at road junctions and vehicles at the junctions positioned themselves by proximity to the IR beams using simple IR detector diodes. By modulating the IR intensity, data was passed from the beacons to the vehicles indicating which beacon was being used and also included vector map fragments of routes to other junctions. (Section 8.3 explains the significance of vector mapping.) In between beacons, a combination of magnetic compass readings, odometer readings, and map matching was used. It is interesting to note that the positioning and navigational side of this experiment performed well without the use of any radio technology. Performance was enhanced by using a large array of IR LEDs mounted behind a lens, which was part of an additional "black" lamp mounted on the conventional three lamp traffic signals.

The first large-scale indoor IR positioning experiment was performed by Want and Hopper [1] with the Active Badge System in 1992 and was aimed, initially, at providing location information for an internal telephone switchboard operator to locate people within a corporate environment. Subsequently the same technology supported many other experiments in ubiquitous computing. The Active Badge was a small battery-powered badge worn by people and contained an IR transmitter that transmitted a code unique to each badge. The rooms of the building were fitted with IR receivers attached to a fixed communications network so that any badges detected within range would be reported to the central control computer that tracked movements. Figure 7.1 shows a simplified block diagram of an IR tag system for tracking people. This approach is the mirror image of the ITS experiment since it is the moving objects that transmit signals and that positioning is centralized rather than autonomous. Section 1.4.4 highlights the advantages of each approach.

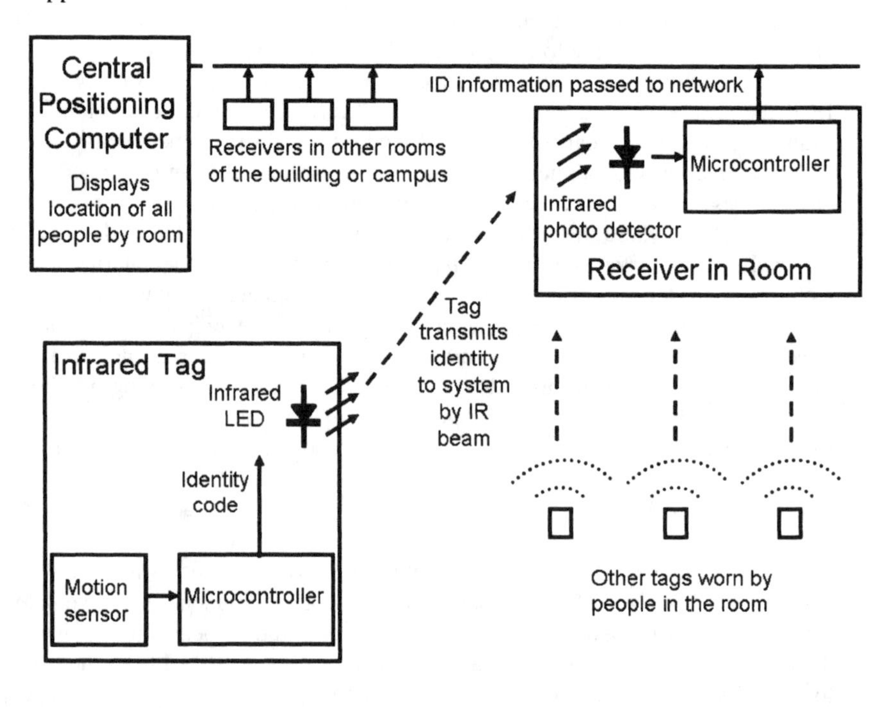

Figure 7.1 Infrared tag personal location system.

The weakness of the IR badge system is that the IR transmitters must always be visible, which is not always easy to manage. A movement sensor was included to detect when badges were not being used (e.g., after office hours) so that battery life could be conserved by disabling the transmitter, but like any active mobile device, battery management is necessary. The great advantage is the very low potential cost of the badges, but on the other hand a dedicated network of room receivers is required. Some experiments have taken place where active badges communicate with each other, leading to the idea of passing location information in a viral manner. Other proposals have been to deploy IR lighthouses where a rotating beam provides angle information (for example, to sectorize a room), but active IR badges have yet to make a major impact on Whereness.

7.2.2 IR Gaming Positioning

The Nintendo Wii games machine uses IR positioning and is currently one of the most popular consumer electronic devices with demand far outstripping sales. A sensor bar is placed horizontally in front of the TV screen with two short rows of LEDs at each end. The handheld games controller contains an IR camera on its end that is pointed in the general direction of the sensor bar. The image of the LEDs is used by the games software to detect the approximate location of each player's controller and is used as an input in various sports simulation games (which are the unique feature of the Wii).

7.2.3 IR Ranging and LIDAR

IR ranging is used in some autofocus cameras using a triangulation or a pulse timing approach but much more accurate ranging is used professionaly in light detection and ranging (LIDAR), which is used in surveying, geodesy, and other scientific activities where accurate distances need to be determined. High-power laser light is used and the longer wave part IR spectrum has the advantage of being eye-safe. Very accurate 3D spatial images can be determined automatically by scanning with rotating beams (using mirrors), and ground terrain models can be made by aerial lidar survey. It is likely that as technology costs fall, advanced techniques such as lidar may enter consumer products and mainstream business, adding a new dimension to photography, art, gaming, amateur mapping, and the surveillance of important physical assets.

7.3 Sonic and Ultrasonic Positioning

Ultrasound is sound that is above the limit of human hearing, which is around 25 kHz for young people and can be used for positioning and imaging. Sound is the longitudinal physical pressure vibrations of matter and travels much more slowly than electromagnetic radiation. In air the propagation speed is around 340m per second (i.e., around a million times slower than light speed) but is an order of magnitude faster in solids. Sonic positioning is used in maritime sound, navigation, and ranging systems (sonar), for applications including finding shoals of fish, the depth of the sea bed, remote vessels, and military targets. On land, the nuisance of loud audible pulses rules out sonar in favor of ultrasonics. The higher frequency of ultrasound also has the advantage of making ranging measurements more accurate; since the wavelengths are much shorter object resolution is easier. Typical consumer products using ultrasound generally operate at 40kHz. In common with IR systems, the consumer components are very cheap and require no licenses. Ranges of tens of meters are similar to IR, and systems are very suitable for indoor use. There are professional applications of ultrasonic ranging, for example, detecting the position of flaws in solid components; imaging ultrasound (operating at much higher frequencies) is now common in medical imaging.

Nature has evolved ultrasonic ranging systems that are used in air by flying bats and in water by dolphins. Bats use their ears and heads to measure angles and use ultrasonics for navigation and finding prey (some of which use countermeasures).

The disadvantages of ultrasonics for positioning are the same as for IR, especially the need for a clear line of sight. A major benefit over IR is, however, the extreme ease with which accurate distance measurements can be made in air. This is possible because the slow pulse flight time allows them to be timed by digital processors that can sample the received pulses at a relatively modest rate well within the capabilities of the cheapest microcontrollers, as the following example illustrates:

Range of object being detected	= 5m
Pulse round-trip flight time	= 30ms
Sample speed of cheap microcontroller	= 10 million samples/s
Pulse resolution (time)	= 0.1μs
Pulse resolution (distance)	= 3mm
Processor samples for pulse round trip	= 3333

It can therefore be appreciated why similar indoor UWB radio systems are expensive and still exotic given the timing circuits and systems are 6 orders of magnitude more exacting!

Ultrasonic signals are easy to produce since all that is required is the production of a 40kHz carrier that can then be connected to a transducer, which is in essence a small loudspeaker consisting of a tuned element made of a ceramic piezo resonator. Receivers are very similar to transmitters and produce an electrical output signal that requires only modest amplification. Since they are tuned resonators their bandwidths are relatively narrow. More sophisticated professional transducers are also available.

7.3.1 Ultrasonic Distance Measurements

A number of consumer products use ultrasonic ranging. Early Polaroid instant cameras used an ultrasonic rangefinder to adjust the camera focus. Today, there are many cheap ultrasonic tape measures available. These send out pulses of ultrasound of increasing intensity until a reliable reflection is received and distance is calculated.

7.3.2 The Cambridge Bats

Active Bats was the name given by a research team [2] at the University of Cambridge led by Hopper for an ultrasonic tag system. It was an indoor positioning system that operated in a way similar to the IR Active Badge systems but with far greater accuracy. The IR system could position only to within the confines of a room whereas the Bats system achieved approximately 3-cm accuracy in three dimensions. The Bats are small tags that contain an ultrasonic transmitter and a telemetry radio receiver.

To achieve the extreme accuracy, a very dense network of ultrasonic receivers is needed that are mounted on the ceiling in a grid pattern of average density 0.6 m^{-2}. A radio signal is used by the central control system to poll each tag in turn that then emits an ultrasonic pulse. Each pulse is received by several ceiling receivers and the flight times calculated. Radio is used only for synchronization and identification, not for positioning per se.

Calculation of position is by circular multilateration and is performed centrally. At least three ceiling receivers must have a distance reading from a given tag for successful positioning to occur (see Figure 7.2). Starting with the first receiver, it can be assumed that the tag will lie on the surface of a sphere centered on the receiver with a radius of the distance measured by the flight time. If a second receiver is then considered, then a second spherical locus centered on that tag will intersect the first sphere. Position will lie on the points of intersection that is a vertical circle (since the ceiling is horizontal). The distance measurement from the third receiver will result in another spherical locus that will intersect the vertical circle at two points: one above the ceiling (which can be ignored since it is an impossible position) and one below, which is the true position of the tag. If full 3D positioning above the ceiling were required, a fourth reading would be needed.

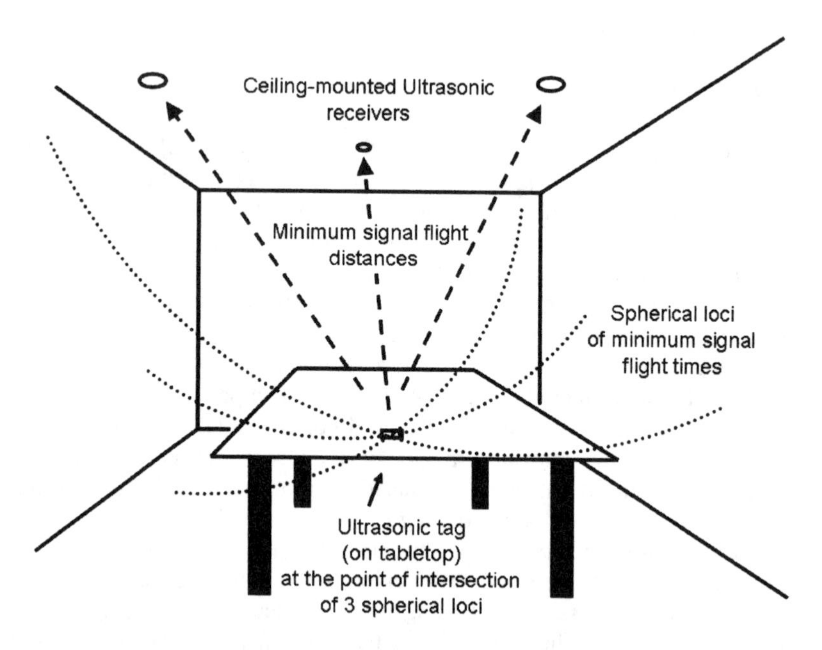

Figure 7.2 Multilateration positioning of ultrasonic tag within a room.

Since many more combinations of readings may be available than just three, statistical positioning algorithms have been developed to produce the best fit of location to all the physical measurements that are available. Repeated measurements will further increase the accuracy of the location. If three tags are attached to an object, it is possible to determine object orientation as well as its location.

Until the invention of UWB radio positioning, the Bat approach has been the best indoor positioning system available. A number of commercial products have been developed and numerous ubiquitous computing experiments performed. Unfortunately, like the more basic IR tag, the system has a number of disadvantages. The centralized approach places a limit to the number of tags that can be used in one environment. While adequate for normal office environments, it would not perform well, for example, in a crowded mass transit situation during an emergency since there would be too many tags needing servicing and a crowded train or station would not allow enough clear paths from passengers using tags to the ceiling detectors. It might be better to work the system in reverse (rather like an indoor version of GPS).

Research work on ultrasonic position is continuing with the current focus being on ad hoc peer-to-peer systems with moving objects being tagged that laterate between each other. Significant progress with the technology could come from the construction industry if building panels were made with tags already embedded or from consumer gaming, taking the approach pioneered in the Nintendo Wii further, with highly accurate indoor 3D positioning and body-wearable ultrasonic sensors. For all its problems, ultrasonics still offers a very good trade-off in terms of accuracy and cheapness.

7.3.3 The Massachusetts Institute of Technology (MIT) Crickets

MIT [3] researched a similar system to the Bats but that worked in a reverse mode. The "cricket" modules were attached to the ceilings of rooms and transmitted ultrasonic pulses. Mobile receivers were carried by people and attached to objects. Before the ultrasonic pulse is transmitted each ceiling cricket transmits a radio signal modulated with identity information. The time difference between the RF and ultrasonic signals is timed and a ranging measurement made. The measurements are repeated by the cricket receiver for a number of transmitters so that triangulation can determine exact position.

The main difference is that this is an autonomous system and unlike the Bat system would not have a specific upper limit on uses that could be supported. A further enhancement (as with the Bat system) is that orientation is possible if more than one receiver is used when attached to an object being positioned. Other sensors may also be incorporated.

7.4 Visual Positioning

7.4.1 Outside-In Positioning

Machine vision is a very effective and emerging technology for finding the 3D position of a target that can be seen or can see. Visual positioning is the primary method that our species uses, and as artificial intelligence and machine-learning techniques advance and become cheaper, the technique is likely to increase. There is an associated reason to use camera technology for positioning: the need for increased security in a more uncertain world. If cameras are needed to enforce and deter crime, then it may be a bonus to be able to enhance the system. First, to recognize targets within a scene; second, to track their motion; and finally, to present the information in a useful and automatic way to those who need it. There are already many users of this technology that include national security systems, computer gaming, and many applications that are relevant to ITS. All the applications where a fixed camera looks into a scene are known as outside-in systems.

7.4.2 Inside-Out Positioning

Another facet of the digital camera revolution is the spread of personal cameras that are now being embedded into other appliances (i.e., mobile phones, binoculars, and laptop PCs). In these cases the positioning is inside-out (i.e., the camera looks out at a scene that will vary as the camera is moved). By comparing what can be seen with what it might be possible to see, it is possible to determine both location and orientation. For example, Cipolla and Robertson [4] have developed a system where a known scene captured previously by a camera can be parameterized (or turned into metadata by encoding the shapes within the scene). This parametric signature of the scene can then be tagged with other information such as location and a description. It is then stored within a library of common scenes (e.g., famous buildings, landscapes, and objects). When a user takes a photograph, a service can be offered to compare its signature with those in the library. If a match is found, location can be found. From the shape and angle differences it is also possible to gain information about relative distance and orientation between the user and the main subject in the scene.

7.4.3 Ubiquitous Digital Cameras

Before looking at some applications and techniques, it is important to know why digital cameras are becoming so ubiquitous and how camera and processing technology is developing. Since the first transistors were made in the early 1950s, it was evident that they were all light-sensitive and measures were taken to keep light away from the active regions by encapsulation. On the other hand, the light-sensitive phenomenon is useful it its own right and a whole range of photo transistors, diodes, and more exotic optoelectronic components have followed.

A computer memory chip is made up of a grid of millions of adjacent transistors. If it is exposed to light, it becomes an electronic version of an eye's retina and all that is needed to turn it into a camera is a simple lens to focus an image. Since memory chips have been falling in price and growing in capability in line with computers in general, it is not surprising that digital cameras have also become very cheap. There are many specialized camera chips: the most basic are CMOS arrays and those of higher performance are charge-coupled devices (CCDs). The continuation of general improvement in silicon chip technology, which includes the cost reduction per bit, an increase of component density, and an increase in the switching bandwidths of the transistors, is known as Moore's law. It is a holistic relationship spotted by one of Intel's founders, Gordon Moore, in the 1960s. It has been true for 50 years and is predicted to continue to hold for at least another decade. We can expect, therefore, that the spread of basic components like simple cameras will continue to the point where anywhere there is activity, there will usually be cameras and most of them will be networked.

Whatever other positioning technology may be available for an application, there is also likely to be cameras close by. Taking their images and using machine vision and data fusion techniques (see Section 7.7), it will then be possible to have an improved Whereness capability.

7.4.4 Vehicle Location

Closed circuit TV cameras (CCTVs) used for traffic surveillance have been in place for many years. More recently, these have been used to automate the detection of individual vehicles based on number plate reading, which is possible by advances in fast computer vision and pattern recognition software. For example, the London Congestion Charging scheme [5], which covers the whole of the most central part of London, uses fixed (and some mobile) video surveillance cameras located at 340 sites. These cameras capture the images of the number plates of moving vehicles within the charging zone. Drivers have to prepay and register their vehicle details, which are then entered within a database. Camera images are processed and the resulting vehicle registration numbers are compared with the database and defaulters fined. Policing has been transformed by the use of automatic vehicle detection that is now used for speed enforcement, detective work, and crime prevention (especially antiterrorism).

Cameras are also used for traffic flow detection. The Trafficmaster [6] system uses roadside cameras to capture random vehicle numbers that are then read again at different locations on the network and the vehicle speed and thus the congestion determined. An RDS-TMC service (see Section 4.1) is offered to motorists equipped with suitable sat-nav equipment, usually via an integrated product included within the vehicle when manufactured.

It is probable that there will always be camera surveillance regardless of other advances in vehicle navigation and automatic driving aids. It seems a lost opportunity, therefore, that the positioning possibilities derived from the roadside camera networks are not being integrated more with other positioning systems, especially those in vehicles based on GNSS.

The ultimate in vehicle positioning is the computer vision systems being used by mobile robots. Competitions are being held between robotic vehicles that are programmed to find their way autonomously across wild terrain and urban environments (mostly for military applications). As the vehicle progresses the stereoscopic computer vision system can reconstruct the scene as a 3D model and then steer the vehicle around a clear path. A similar approach is being taken with robotic rover vehicles exploring the surface of the planet Mars, and others are planned. Computer vision systems with scene modeling may be used for consumer driving assistance in the future, although it seems unlikely that the driver would be entirely robotic. Stereo vision systems are also used in digital navigational road mapmaking (see Figure 8.1).

7.4.5 Motion Capture

Very accurate camera-based 3D positioning is used in motion capture systems employed by animators in video content production. For example, realistic human motion from a real actor is captured so that the motion may be modeled for use with an entirely artificially generated computer-based character (modeled in 3D graphics software). The character Gollum in the *Lord of the Rings* trilogy was created in this way. As well as body motion it is common to capture detailed facial movement in order to generate realistic lip movement for dubbed voiceover.

Various techniques are used including the positioning of passive reflective dots at key motion points and active LED beacons. These techniques are very tightly bounded and careful camera positioning and calibration are essential. Large arrays of cameras may be used, and it is still an expensive and highly professional activity. It may however, become possible to offer consumer variants especially as home computing capabilities (especially multicell graphics processors) improve. A future games machine could include motion capture positioning perhaps from several small cameras to cover a wide area to enable gamers (or artists) to interact and act along with screen-based, computer-generated characters. This is one of the ways in which the real world and its mapping will become enhanced with overlays from virtual worlds leading to augmented reality (see Section 5.5).

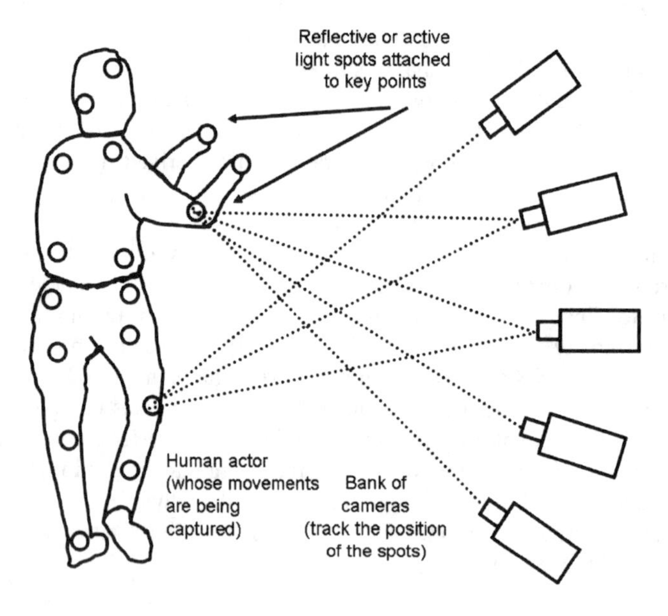

Figure 7.3 Motion capture positioning using a bank of optical cameras.

7.5 Movement and Inertial Sensing

Dynamics is the branch of physics that covers the motion of objects and includes the variables of position, distance, velocity, and acceleration. These are all related by the laws of motion or Newton's laws, which describe the movements of all objects and are accurate for all but the most extreme situations when Einstein's theory of relativity also needs to be included. Computer navigational systems employ a "motion model" using a set of algorithms based on these mathematical relationships (together with some calculus, which Newton also invented). The model attempts to predict, using software, the real physical motions of objects. Inputs to the models come from sensing, measurements, and a priori information and the model is only as good as the underlying quality of the sensed information and the degree of sophistication. It is possible, however, to improve quality by using statistical filtering techniques, but the computational load of modeling and filtering can be immense. One of the main reasons why Whereness is an emerging opportunity is that is it dependent on improvements in computer science.

We have considered systems that measure distance by timing pulses and geometry and velocity (i.e., speed in a given direction), but now we come to linear acceleration and rotation. Acceleration is very easy to measure, especially with the advent of accelerometer chips and rotation also with solid-state gyroscopes, but the relevance of acceleration needs perhaps a little explanation.

Acceleration is the rate of change (with respect to time) of velocity and in turn, velocity is the rate of change of distance. So it can be seen that if timed measurements of distance, velocity, or acceleration are taken, the others can be, in theory, determined. In mathematics, velocity is the first derivative of distance and acceleration its second derivative. If rate of change of distance is differentiated, velocity is obtained and again, if velocity is differentiated then acceleration is found. Conversely (and of interest if we wish to use accelerometers), if the output of an accelerometer is integrated with respect to time, we obtain velocity and if velocity is integrated distance is found. These relationships were first computed using analog computers that are adept at calculus and that guided the first generations of electrically controlled weapons, but today we use digital algorithms and processors that can model motion in real time.

Accelerometers can also be used to determine tilt by measuring the effects of gravity as well as inertia. Gyroscopes (gyros) are used to detect turning moments and direction of travel. Historically they were based on rotating flywheels and were and continue to be used in aircraft and missile autopilots. There are now families of solid-state miniature devices based on other techniques, (e.g., quartz crystal tuning forks). These mimic halteres, which are nature's gyros and which can be observed on the bodies of some insects such as the crane fly. As these structures vibrate they tend to oppose rotation that can be detected by changes in their oscillations. Along with accelerometers, gyros are now in consumer electronics applications such as radio-controlled model aircraft and vehicle navigation systems.

Inertial systems are beginning to have a major impact on consumer electronics as the original military technology is adapted now that the basic components are so cheap. Accelerometers (and other sensors) will be increasingly finding their way into more and more devices like mobile phones, cameras, gaming machines, toys, and models, and anything with a human machine interface (HMI). All these devices will now have the potential to find position by autonomous inertial methods, and the impact on Whereness will be profound.

7.5.1 Vehicle Positioning Using Wheel Sensing

Vehicle motion has traditionally been sensed by methods based on wheel rotation, which gives an accurate measure of distance provided there is no wheel spin, skidding, or sliding. Mechanical systems took a rotational feed from the vehicle power train (after the gearbox so that that rotation was a constant ratio with respect to the wheels) and provided the driver with a mechanical (or electromechanical) speedometer and odometer. Some mapmaking systems have used vehicles fitted with a fifth wheel centrally mounted at the rear of the vehicles. They provide more accuracy because they eliminate the effects of differential wheel rotation when turning. However, the differential effect can be useful for navigation since it provides turning information. The ancient Chinese south-pointing carriage [7] was a two-wheeled towed vehicle that used a differential gear to show when a straight course had deviations. If one wheel rotated slightly more than the other, then a curve was being followed and the indicator pointed the need for and direction of a course correction.

The modern equivalent is the use of ABS pulses from each rear wheel. Some antilocking breaking systems (ABSs) are electronic and use a system where the rotation of the wheel is sensed magnetically and a series of electronic pulses proportional to wheel rotation are generated. Differences in pulse rates between the rear wheels allow a processor to calculate accurate turning information. Some vehicle navigational systems used these pulses with map matching as an alternative to GPS positioning (especially before GPS was available). The advantage of autonomous navigation is that it relies on no external signals, unlike GPS systems (see Section 6.2.3).

7.5.2 Pedometers

Pedometers are devices that sense and measure footsteps. They are the human equivalent of vehicle odometers but as legs do not rotate, motion is difficult to measure accurately. It is not usual to measure the length of a footstep directly so an indirect method is adopted where the upward and downward motion of the hip is detected instead.

A simple mechanical pedometer can be made using a small horizontal pendulum that is spring-biased to be effectively weightless in the vertical plane.

The devices are worn at the hip and when it rises and falls during walking, the pendulum (which by inertia tends to stay motionless) touches a switch contact that in turn feeds an electronic processor. Averaged footfall count, speed, and other parameters are calculated (see Figure 7.4). The system needs careful calibration since it works on the basis of average length of footstep. Turning corners, shuffling, or climbing all lead to errors so the devices are quite inaccurate (5% is typical). A more sophisticated approach uses an accelerometer chip in place of the pendulum.

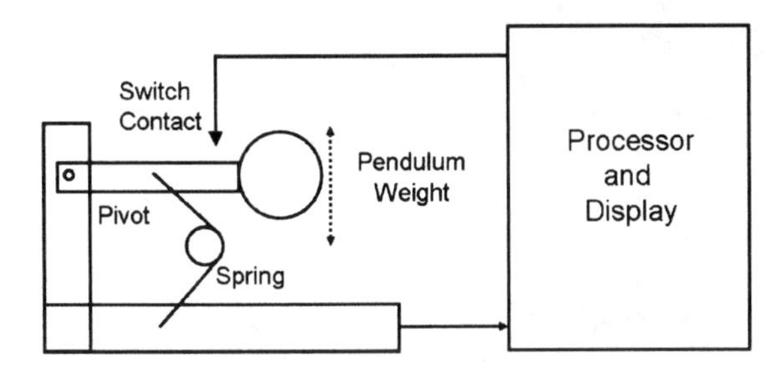

Figure 7.4 Mechanical pedometer.

As the hip rises and falls, acceleration is greatest as it changes direction, and the accelerometer will therefore give a peak reading. It is normal to filter the pulses to avoid any anomalies. Accelerometer chips have the advantage of no moving parts that are subject to friction and wear and are very small. They can also be used for other useful purposes, for example, to detect the orientation of the wearer by sensing the force of gravity on up to three axes. An injured person who is on the ground, for example, would be easy to detect since the footsteps would have stopped and the orientation of the device would be different. We now consider how these inertial chips function.

Figure 7.5 A typical three-axis accelerometer chip and solid-state gyro module.

7.5.3 Accelerometers and Gyroscopes

Accelerometers and gyroscopes take many forms, some of which are used only in aerospace and military applications. Accelerometers function by measuring the displacement of a mass due to gravity or inertia but it is the recent availability of highly miniature devices based on MEMS technology that is having the most day-to-day impact. Microelectromechanical systems (or MEMS) are devices that are nonelectronic components constructed using silicon chip fabrication processes. For example, a simple flexible cantilever will act as a one-dimensional accelerometer. If it is mounted on top of a silicon chip, the flexing of the beam can affect the capacitance of an integrated capacitor, which in turn can alter the frequency of an integrated electronic oscillator, which can then have its frequency counted in a processor subsystem and so on until the acceleration information is in the desired format for navigation or some other application. If three similar cantilevers are made and mounted in three planes, three-axis acceleration can be measured and a full three-dimensional motion model can be used.

When following a route a gyroscope is useful to provide headings and is not the subject of external influence (as would be the case were a simple magnetic compass used). Both accelerometers and gyroscopes do not provide absolute references so there is a need to link their relative reference frames to that used in any associated map and the real world. For this reason both gyros and accelerometers are often combined with other positioning devices. For example, a gyro can be calibrated versus a magnetic compass and will then be useful to show when the compass is indicating anomalies. Likewise, accelerometers suffer from drifting errors especially during slow motion, so they can be calibrated against a radio-based system that may be intermittent.

It is usual, therefore, to combine GPS systems with inertial systems. The former can provide an absolute position but with outages due, for example, to poor sky visibility, and the latter can fill in the coverage gaps.

7.5.4 Smart Floors

Many research projects [8] have been experimenting with the idea of a smart floor that can detect footfall and track the motion of people walking and dancing. The smart floor is really a scaled-up version of the touch screen found on many computer interfaces and they are potentially very accurate. Pressure sensors or other transducers based on, for example, capacitance changes (see Section 7.6.2) are mounted in a dense grid under a flooring surface so that the presence of feet can be accurately located. The computational challenge is to follow the progress of several people across a common area and to find methods to identify them. One approach would be to use some sort of door access system to open the door to known people. At the threshold, it is clear to the tracking computer to whom the feet belong, so that the individual can then be tracked unambiguously. Anomalies are possible if people jump, are carried, or if too many feet are in a very small area, but the technique has many merits.

Perhaps the main attraction is that users do not need to carry any equipment or to have any training. The system is invisible and it is possible (although currently expensive) to make dedicated active flooring. New nanotechnology materials are emerging that may be useful, for example, quantum tunneling compound rubber. This conductive rubber is highly sensitive to pressure since it is impregnated with nanoparticles of metal that when compressed into close proximity, start to conduct electrons using the quantum tunneling effect.

In common with much research in ubiquitous computing, smart floors (and smart furniture) are still emerging as products but are likely to have a major impact on the future built environment. Already there are new art forms based on dance and music seeking to exploit smart floors, which are effectively part of a musical instrument, a computer interface, and a positioning system (see Section 5.7.1).

7.6 Positioning with Magnetic and Electric Fields

7.6.1 Magnetic Compasses

Since antiquity, positioning due to the Earth's magnetic field using permanent magnet compasses has been a fundamental part of navigation. Today, small and inexpensive magnetic field sensor chips are included in consumer products and professional compasses are still used if only as a standby for when all else fails. The electronic chips are based on thin magnetic films that change their electrical properties when exposed to magnetic fields and are small enough to be incorporated into conventional wristwatches. Some pre-GPS vehicle navigation systems used electronic compasses, and electronic compasses are still a popular vehicle accessory.

The main disadvantages of all compasses are due to the erratic nature of the field. Iron structures, electrical plant, and northerly latitudes (where the horizontal field component is weak) all cause problems. A further longer-term concern is the gradual weakening of the field over the past decades, which may be a precursor to a field reversal during which time the field is likely to be unusable. Each year the magnetic poles move, and this is another factor that has to be taken into account.

Compasses will not disappear and the trend may well start to increase when new applications are exploited as basic positioning creeps downmarket and starts to be included in new generations of consumer devices, especially toys and games.

7.6.2 Electric Field Proximity

Positioning of objects by close physical proximity, using electric fields, is a technique that is increasing. For example, a touch pad or touch screen can detect the close proximity of a hand or finger. In a vehicle, an infant seat can detect the presence of a baby. Oscillating electric field devices are used to detect the changes in capacitance that occur when many objects are moved within a short distance of a sensor, which is usually a simple pair of conductive plates.

The advantage of these techniques is that no moving parts are needed since actual touch is not needed. The latest device to exploit this technique is the screen of the Apple iPhone. It is likely that the use of active surfaces will increase and together with active flooring will be part of smart buildings and furniture of the future.

7.6.3 Oscillating Magnetic Field Positioning

Digitizing tablets are used as peripheral devices to input graphical and map information into computer applications (see Chapter 8). These tablets are flat surfaces with embedded magnetic field windings (i.e., coils of wire) below the surface. A moving puck or pen is used by the operator and it is also fitted with an embedded winding. They operate using a magnetic field that is generated by the windings connected to an electronic oscillator. As the changing field propagates across the surface, interaction with the moveable coil (in the pointing device) is detected because the subsurface embedded coil acts as both a transmitter and receiver of magnetic energy. The moving coil can either be passive or active depending on the exact design, and a degree of orientation detection is also possible. Cartesian position is outputted by the sensing system and precision is very high (this is the most accurate positioning system discussed in this book).

Room-sized experimental versions of this technology have been used for detailed indoor positioning experiments. The ideal situation is to use coils that are located at the edge of the space, but the problem is that the power needed to generate a field over extended distances is large and the coils inconveniently large. It is also inconvenient to use a mesh of coils spread out under the floor. Unlike the pressure-sensitive floor, the subject being tracked must carry a coil that interacts with the field. The overall approach may be more reliable than pressure-sensitive flooring since there are no moving parts to wear but at present the technology is cumbersome.

A further use of propagating magnetic fields is in underground communications and positioning. Although something of a niche market, cavers and those working underground can use specialist magnetic field equipment to both communicate and navigate. Magnetic fields will propagate well in environments where normal radio will not penetrate but the disadvantage is that range and bandwidth are limited and power requirements large. It is, however, possible to send voice communications between the surface and subterranean spaces and using directional coils to detect the direction of the signals.

7.7 Sensor Fusion

Multisensor data fusion is a technique used in many areas, including medical diagnosis, battlefield target recognition, and the control of autonomous vehicles. It is based on the idea that, although it is useful to process a single source of data using statistical techniques, the use of multiple data sources "fused" together can sometimes significantly improve the accuracy of measurements. This is particularly useful where one inadequate source of data can be compensated by another. We have already seen several examples in this book where a multisensing situation is better than with a single system, including:

- GPS and cellular radio;
- GPS and wheel sensors;
- GPS and WiFi Indoor Positioning;
- Personal autonomous navigation (pedometer, compass, accelerometers, gyros);
- Map matching and GPS.

Fusion techniques for general sensors are well documented but often from a military perspective [9], and it is only recently that fusion is beginning to make an impact on consumer and business applications. Hightower et al. [10] introduce the concept of a "location (protocol) stack" that includes sensor fusion for location. Böhringer [11] describes a novel system that was used accurately to position trains and is a good example of fusion at work. Data from a radio positioning system (GPS) is fused with that from a magnetic field sensor attached to the train wheels (which can pick up characteristic magnetic signatures from track joints and equipment). Clearly, GPS will not work in rail tunnels but rail joints are always present and provide complementary information. To extrapolate between readings, mapping data is used that can be considered to be a third source of information to be fused. Map matching is an important part of navigational data fusion.

The algorithms used are complex but a common approach is to use variants of the Kalman filter (see the Epilogue for further details). It uses an estimation process that is updated based on real measurements and an ongoing motion model.

It can be seen from both this chapter and Chapter 6 that there are many ways to find position and if Whereness is to be achieved, there must be methods to usefully combine them. Multisensor data fusion is the key low-level technique needed to deal with measurements and detailed mapped information (which are really a priori averaged measurements). The problem at the moment is that there is a wealth of information that can be sensed, measured, fused, stored, mapped, and shared but no universal way to fit the "jigsaw puzzle" together. This is a situation not unlike the more general information maze that is available on the Web. We know the information is there but its sources and its formats are not yet well structured. This situation is changing with today's Web 2.0 and the Semantic Web that is under development. In the next chapter, we will see the fundamental importance of Web 2.0 methodologies as they are applied to maps and in the final chapter, we will consider, in the future, how positioning systems could automatically help determine what can be fused using a Semantic Web approach.

7.8 Summary

This chapter covers a range of different sensor and sensing technologies that are not radio-based but that can still be very useful to find position. The advantage of these is that they are generally very simple and cheap solutions and since they do not use a licensed propagation medium, are license-free. The main disadvantage from optical and sonic positioning is the need for an unobstructed line of sight.

Optical systems can use either infrared communications usually by simple proximity or visible light and employ digital cameras. Cameras are being increasingly used for surveillance and the addition of machine-learning software means that it is now possible to recognize and track targets within a scene. This is the outside-in approach. If, however, mobile cameras are used (inside-out), they can be used by mobile users to capture images of landscapes, skylines, buildings, and objects that can be compared with a known library of images to provide positional information. The most accurate positioning systems are optical and are used for motion capture by cinematic animators. Very complex camera arrays and software are used to track the detailed motion of key parts of the human body.

Ultrasonic positioning is very accurate (similar to UWB radio). Many successful research experiments have been performed, but the downside is the need to install a very high density of sensing units within a building, which can be expensive.

Inertial and mechanical sensors are useful since they do not need to use any external medium. Measurement of acceleration by a new generation of MEMS chips is adding a positioning capability to many consumer appliances. Solid-state gyros, wheel sensors, and pedometers are also used. Infrastructure such as flooring can also include movement sensors using either mechanical or field sensing.

All this sensing capability together with the radio methods can be combined by the technique of sensor fusion. It is a key element in the convergence needed to deliver Whereness. If maps are also included within the fusion algorithms performance is improved further.

References

[1] Want, R., Hopper, A., Falcao, V., Gibbons, J., Olivetti Research Ltd. 1992 "The Active Badge Location System," http://citeseer.ist.psu.edu/146434.html, Jan. 2008.

[2] Harle, R., Beresford, A., "The Bat System: Ubiquitous computing in action." http://www.vs.inf.ethz.ch/events/dag2002/program/ws/Beresford-Harle.pdf, Jan. 2008.

[3] *The Cricket Indoor Location System*, http://cricket.csail.mit.edu/#overview, Jan. 2008.

[4] Randerson, J., "Lost? Send Snap and SOS to," *New Scientist*, 10 Apr. 2004, p. 24.

[5] *London's Congestion Charging Scheme*, http://www.cclondon.com/index.shtml, Jan. 2008.

[6] *Trafficmaster*,http://www.trafficmaster.co.uk/about_trafficmaster/about_us.php, Jan. 2008.

[7] Drane, C.R., Rizos, C., *Positioning Systems in Intelligent Transportation Systems*, Norwood, MA: Artech House, Boston, 1998, pp. 2-3.

[8] Orr, R.J., Abowd, G.D., "The Smart Floor: A Mechanism for Natural User Identification and Tracking, Graphics," Visualisation and Usability (GVU) Centre, Georgia Institute of Technology, http://www.cc.gatech.edu/fce/pubs/floor-short.pdf, Jan. 2008.

[9] Waltz, E., Llinas, J., *Multisensor Data Fusion*, Norwood, MA: Artech House, 1990.

[10] Hightower, J., Fox, D., Borriello, G., "The LocationStack, " http://portolano.cs.washington.edu/projects/location/, Jan. 2008.

[11] Böhringer, F., "Train Location Based on Fusion Satellite and Train-borne Sensor Data," *Location Services and Navigation Technologies*, Proc. SPIE Vol. 5084, Apr. 2003.

Chapter 8

Maps and Whereness

8.1 Why Maps are Important

Of all the areas discussed in this book, this chapter is the one where the technology and practice are developing most quickly. It cannot be overstated how important the advance of new World Wide Web (WWW) technology is to maps and Whereness. Currently, the somewhat imprecise term "Web 2.0" [1] has been coined to describe the movement that is taking the original Web concept of one-way browsing content into a collaboration model where there is a sharing process with users posting content and organizations allowing their content to be automatically manipulated by users and third parties.

At the start of the book, in Section 1.4.1, the historic importance of maps in navigation was highlighted. The synergy between positioning and mapping has been increasing and will continue in the future as traditional maps take on new digital forms and they then help inform other electronic and information systems.

This chapter will therefore describe both mapping worlds: the traditional digital world, which uses specialist GIS software used by trained experts and the new Web 2.0 methodology sometimes described as the Geoweb, which is increasingly touching the lives of everyone via intuitive Web interfaces, such as Google Maps. It will also be describing how digital maps of all descriptions are changing from being a two-dimensional representation of the Earth's surface into something more helpful by adding height and becoming more of a 3D pictorial format. Maps are starting to include the fourth dimension of time, so that former unchanging (or static) maps now include dynamic or changing information that is often provided in real time, for example, by positioning or sensing systems described earlier in this book. *Traffic England* [2] is a good example (see Section 8.8).

The way positioning systems can help build maps will be described and in particular the new approach of open collaborative mapping, where private people are building maps that may be more up-to-date and accurate than the traditional commercial digital maps. This approach is a potentially disruptive technology for the established players (as highlighted in Section 2.1.2 and in Chapter 3).

If the goal of ubiquitous positioning is to be achieved, there are huge gaps in mapping that must be filled. Although the outdoor environment is very well covered in the developed world, the maps do not extend indoors or underground. There usually are maps of these areas but they are usually referred to as plans and are drawn as part of the construction process and not generally available to users. It is hoped that as developing nations develop economically their maps will also improve especially as remote sensing techniques are making traditional mapping less expensive and more convenient. There is, however, a good prospect that the open collaborative approach might just fill all the mapping gaps and lead to a world where everywhere that people go is accurately mapped. The Whereness vision will then have been achieved. The final chapter will discuss how truly universal mapping and positioning may come about.

8.2 Using Web 2.0 for Maps Today

The current speed of change in the business side of digital mapping will inevitably make any "snapshot" description out-of-date. More generally, many of the pioneers of Web 2.0 are being taken over by bigger players and it looks as if there will be an emerging ecosystem of a few very big players, each one of which will have a mapping arm. At the time of writing (Dec. 2007), the state of the industry in no particular order is outlined in the following sections.

8.2.1 Google

Google started as a search engine but is becoming a more general-purpose information provider covering many useful domains of knowledge. Its two mapping offerings are some of the most popular Web 2.0 sites. Google Maps [3] offers seamless global 2D mapping with street-level details and extra user selected facilities such as route planning and overlays showing satellite images and terrain. It integrates maps from other parties (e.g., Navteq and Teleatlas) and provides an application programming interface to allow users to combine and mashup other data with Google Maps to create other applications. "Street View" is available in popular locations and allows interactive photographic views of streets to be manipulated.

Google Earth [4] is similar but delivers digital satellite images and aerial photographs, some of which have been made to appear somewhat three-dimensional and can include 3D models of cityscapes. (Type "New York City" into Google Earth for a stunning view of Manhattan.) Google has stated it is interested in organizing the world's information geographically, using the map and earth/global as familiar interfaces to all types of information.

8.2.2 Microsoft and Multimap

Virtual Earth is Microsoft's on-line product (which is similar to Google Maps), and MapPoint is a stand-alone enterprise GIS application. Microsoft recently took over Multimap [5], the popular Web 2.0 mapping company [6].

8.2.3 Teleatlas and Tom Tom

Tom Tom [7] is a popular manufacturer of vehicle satellite navigation (sat-nav) equipment that has recently taken over Teleatlas [8] (a provider of vector maps for navigation for many other parties, including Google).

8.2.4 Navteq and Nokia

Navteq [8] is a competitor to Teleatlas [9] and has recently been taken over by Nokia, the largest manufacturer of mobile phones globally.

8.2.5 Mapquest and AOL

Mapquest [10] was one of the earliest entries and a very popular site for street-level mapping that is now part of AOL (a large international Internet service and content provider), which is in turn part of the media company Time Warner.

8.2.6 Other Interesting Services

Other interesting services are provided by ULocate [11], a service for mobile phones, AboutMyPlace [12] for very detailed aerial photography of the U.K., and ViaMichelin [13] for travel and tourism.

8.2.7 The Social Networking Phenomenon

It seems likely that as the current trend for Web 2.0 social networking increases as sites such as Facebook, Bebo, and MySpace become the first Web port of call for an increasing section of the population, these organizations will also start to have a mapping aspect to describe locations of the subjects and their activities. If these are combined with real-time information (for example, from mobile phone activity), then a good part of the Whereness vision might be realized. Currently, many are questioning the long-term effect of social networking and its market value, but there is a good chance that this list of big mapping players will soon include aspects of social networking.

Social networking may soon spread to include business and commercial networking and be the de facto method (at least in some business areas) to communicate with colleagues, partners, stakeholders, suppliers, and customers. The inclusion of Whereness services could be promoted by the clear business benefits (as outlined in Section 4.9). Some services such as Meetro are already operational [14].

8.3 Some Digital Mapping Basics

8.3.1 Maps as an Interface

When one is considering digital maps there are two aspects to keep in mind; first, the map that is seen and used via a human interface, and second, underlying information that will usually be contained within a well-defined data format. Both are described as maps. The image of a map seen on a computer screen is created by an array of colored dots (or pixels) and is similar to the same map when printed on paper. These are know as raster images and are an important part of computer multimedia but are there primarily for the benefit of people. Images can be superimposed easily and points of interest (PoI) can be plotted on the map to help find facilities.

8.3.2 Invisible Maps

The mapping metaphor has been taken over by computer scientists to refer to the activity of structuring any data conforming to some sort of reality, but in this case we are considering the mapping of maps (i.e., structuring geospatial information).

The drawbacks of computer images are that they are inefficient to store and manipulate and do not form an "information space." Information spaces in computer science refer to an abstract view of structured data that is often contained within a database. If the information is coded within a normalized structure, it may be used by software directly.

Machine-learning techniques are being found to convert pictures into more structured formats: examples include automatically picking out roads from an aerial photograph and turning a vehicle number plate image into plain text (to be inserted automatically into the congestion charging ticket!). Often the process of structuring is done manually, for example, with a digitizing tablet. Once captured the information is described in a parametric form (or metadata). A straight line on a 2D map could be described by two sets of x-y coordinates and a curved line either by a set of coordinates or as a mathematical function seeded by a subset of points (see the Epilogue for a more detailed explanation). This sort of map is known as a vector map. In mathematics, a quantity with a size (or magnitude) and a direction is defined as a vector. Most digital maps therefore have an underlying vector format with a rendering process that is used to take a required subset of the vectors and to render it into a raster image that can be displayed or printed.

8.3.3 Positioning on a Map

If a position of an object or person is known from some sensing system, it can be mapped. The physical sensing process will result in measurements and the creation of values (or numbers) that can then be transformed by an algorithm into coordinates according to the relevant reference frame. Sets of coordinates are then added as new vectors that can be displayed if required (in the raster image). It can be appreciated, therefore, that with each new set of measurements the information space increases in size and usefulness. Not only can current positions be shown, if time stamped, but past positions and tracks can also be displayed. Most positioning systems that capture motion will store track-logs as sets of coordinates with time-stamps. Future positions can also be inferred by using algorithms to predict motion or improve accuracy by smoothing out sensing errors. (The Epilogue contains details of the Kalman filter and Particle filter that are commonly used.)

8.3.4 Geographical Information Systems (GIS)

Since the early 1980s the GIS industry has emerged, and very powerful software applications have been written that depend on powerful and massive underlying databases. ARC/INFO from ESRI was one of the first systems, but today there are many (e.g., Oracle Spatial). Industry standardization groups, in particular the Open Geospatial Consortium (OGC), have been established.

GIS can be used in a number of ways. A GIS desktop application may be used by a trained expert to create customized maps, for example, the location of a company's best customers, and be combined with a tracking system to show the progress of a truck en route to them. It may be used as a tool, for example, to plan optimum routes or to store historic information and be linked to other enterprise systems. Sometimes a GIS may be invisible (i.e., the underlying information space is used but no conventional map might be displayed). A reverse emergency call service might use a GIS to work out what proactive calls and messages to generate (see Section 4.5).

The term "distributed GIS" is used to differentiate between the original single-user GIS that would be used on a single work station and the situation today, where generally the components are not colocated. Enterprise GIS is used within an organization and is usually distributed and has licensed mapping content—an example is Microsoft's MapPoint. A Web 2.0 approach will (by definition) be distributed and may serve an enterprise or the general public. The content will be covered by a licensing agreement with the mapping owner but the service might be free for personal use (with perhaps advertisements providing a revenue stream to pay for the content). An exception is, however, the growing open mapping movement where the mapped content is an "information common" (see Section 3.3) and will be described in more detail in Section 8.7.

8.3.5 National-Scale GIS

Mapping agencies used advanced and networked GIS to sell digital maps on demand. Mapmakers might use a digital landscape model as the underlying information space with a cartographic model created on demand to fit specific visual requirements. For example, the Ordnance Survey's (OS) national geographic database model for Great Britain (population 60 million) consists of around 500 million entities, each of which is uniquely identified by a 16-figure topographic identifier (known as a TOID). There are around 5000 changes per day to this massive vector database.

GIS are constructed as an information hierarchy. The foundations are the spatial vectors forming referenced shapes or polygons. These polygons will then have attributes to show various themes (road, rail, water, parkland, farmland, contours, etc.). Although the map is essentially seamless, it will need to be tiled to cover useful areas. There may be other referencing systems associated with the map, for example, postal code areas and the addresses of buildings and cross references to information spaces held by other organizations. National GIS are controlled by governments and their agencies and are of strategic significance both economically and militarily. The obscure name for the Ordnance Survey illustrates its military origins in 1791.

Today, the importance of national mapping is growing as it provides the raw material for the availability of Web-based maps offered via Web services by the various Web 2.0 map service providers and vehicle navigation mapping organizations.

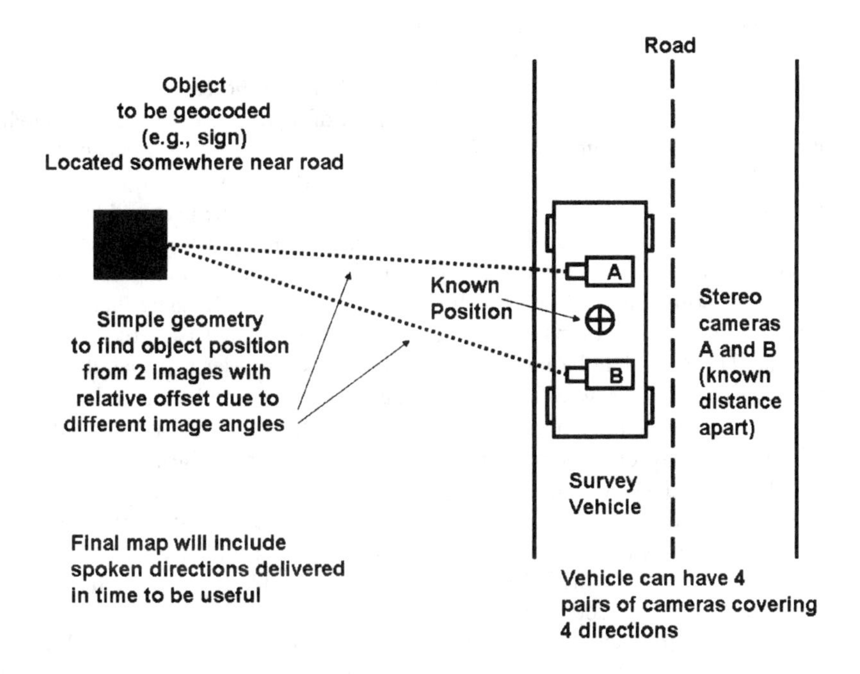

Figure 8.1 Finding the position of roadside objects using stereo cameras.

Organizations such as Navteq and Teleatlas buy licenses for the raw road vector data from national agencies and then add extra layers of information for use by drivers. These include attributes such as turn-by-turn instructions and point-of-interest information. Camera vans are used to make video surveys (using stereoscopic ranging) so that exact GPS locations of the van can be used to find the location of extra attributes that can be seen in the distance away from the road (see Figure 8.1). Manual attribution is often necessary and is a time-consuming process often completed where labor is cheap. This is a good example of positioning by inside-out scene analysis.

Rob van Essen [15] describes maps evolving from mathematical line graphs to virtual reality models and [16] describes the Tele Atlas mobile van technology with six cameras (two forward-looking stereo cameras for geometrically accurate measurements from captured images and a sub-0.5m positional accuracy with pictures of 1300 × 1300 pixels resolution). They are also experimenting with 3D using a gyro for capturing road gradients and laser scanners for transverse slopes (i.e., banks) and street layouts and with other combinations of sensors to provide additional information using sensor fusion (as described in Chapter 7).

8.3.6 The Geoweb, Web 2.0, and AJAX

Simple raster image maps have been available since the early days of the Web. When a request was made by a user for a map of a particular sort, the GIS behind the server would create a raster image file (or bring it out of a cache if already created) and serve it as a simple image. While this approach is still used, the idea behind Web 2.0 is more complex but far more useful to the user and less wasteful of computing resources at the host. Rather than sending a single raster map, the modern approach is to send metadata from which the required map may be created within the software associated with the user's browser. Metadata includes scripts that are similar to computer code but is usually in an easily readable textual format. The script can be interpreted locally in the browser and may then interact with the normal browser's user controls and other media served within the metadata. Thus a map may be zoomed, panned, or otherwise changed locally without the need for the server to keep generating new map tiles.

The metadata is also available via an application programmable interface (API) so that third parties can use it as part of a more extensive application. The term "mashup" is used for this approach of combining information from different providers. Access to the API may be free but normally requires the use of a registered electronic code key.

Maps are thus served in chunks with everything the user might need included but only displayed when required under control of the script and controls. The current name for this approach is asymmetric javascript and XML (AJAX) [17]. The term "asymmetric" covers the idea that the sessions are infrequent compared with user interaction. "Javascript" is a standard used within many Web applications that is similar in syntax to the common Java programming language. The Javascript is not the same as a Java applet, however. An applet is a small program that runs locally within a controlled environment whereas the script controls entities already present within the browser environment. Figure 8.2 illustrates AJAX and its efficiency advantages over the original approach.

The concept of running a local map application is particularly useful for mobile users with mobile Internet connections and local positioning systems that may sporadically connect to the server. Fragments of map may thus be downloaded (triggered by location) but used continuously with local real-time sensed information overlaid.

8.3.7 Tips and Tools for Electronic Cartography

Mapping Hacks [18] is an excellent book for the technically minded that explains electronic cartography very well and has 100 worked examples of digital mapping applications that can be followed with moderate computing skills. It covers mapping of one's life, neighborhood, the world, the Web, and mapping with gadgets such as GPS receivers, desktop applications, names and places, other people, and the Geoweb.

The use of the term "hacks" is benign and does not imply criminal hacking but rather an informal approach to rapid prototyping that (amateur and professional) computer scientists enjoy.

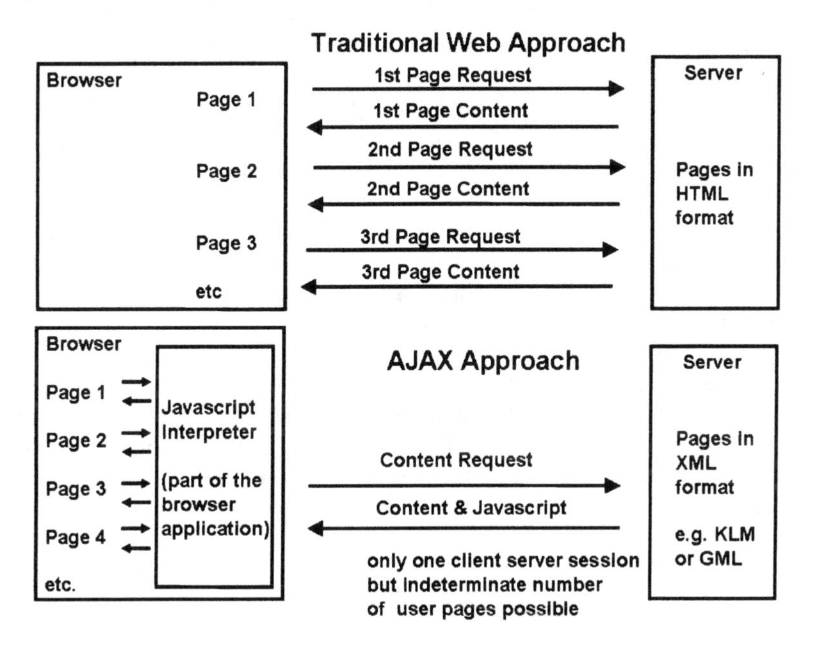

Figure 8.2 Web client-server operation and AJAX.

An important aspect to Web 2.0 is the ease and rapidity with which new innovations can be created. In the past, complex software has been very time consuming and expensive whereas with the mashup approach the process is becoming somewhat automated. Hacking books (of which there are many) are enabling moderately expert programmers to make huge strides and to spread advanced ICT techniques into many domains. As businesses exploit these amateur activities, great economic rewards are likely to follow, hence the interest of business journals such as *The Economist.*

8.4 Web 2.0 and Whereness

8.4.1 The Geoweb

"The World on Your Desktop" [19] is an excellent overview of the emerging Geoweb; published in *The Economist*. It begins with a definition, "As the Internet becomes intertwined with the real world, the resulting Geoweb has many uses." It covers a host of current Web 2.0 applications useful to people and businesses and tricky issues such as individual spatial privacy. It concludes with a vision of "extrasensory information awareness" where mapping personal information is taken to extremes and looks to the next stage of the Web (which is covered in the next chapter). Perhaps it is noteworthy that Whereness as a concept is now mainstream to the extent that it is featured in a conservative business journal.

8.4.2 Google Maps and Hacks

Google has made the biggest impact to date with its global mapping coverage and easy interface. Commercial versions are appearing embedded within other commercial systems as Google's business model expands from an advertising-based model to a license-based one. Simple tools are available to personalize maps with extra information (e.g., photographs).

Perhaps the most exciting aspect for anyone technically minded is the ease by which new applications can be made via the API. Several books are available that give worked examples of applications. For example, [20] shows how to incorporate Google Maps within a personal Web site, shows various tracking and transport applications, dealing with pictures, creating a geo-blog, adding custom maps, and all manner of mashups.

Associated with Google Maps is Google Earth, which is a more visually compelling Web 2.0 application that has stitched together thousands of digital aerial photographs or images (taken from aircraft and from space). In certain built environments (e.g., New York City) it includes a 3D model of the buildings too. Virtual tourism is thus possible from any browser with a broadband connection. Unlike Google Maps, which is an AJAX system, Google Earth requires a local application to be installed that is required to process the imagery (a task currently beyond the relatively limited capabilities of Javascript).

At some time in the future it seems likely that this application or something similar will start to include real-time imagery, sensing, and positioning technology.

8.4.3 3D Building Models and Virtual Worlds

Google has provided Sketch Up, which is a toolkit PC drawing application for people to produce 3D models of buildings (or anything else) that can be added to Google Earth. This may be the way that mapping gets extended into the areas not currently mapped. Building plans could be used as a starting point for the creation of the Sketch Up models that can be made to scale and that tessellate with outdoor mapping.

3D modeling is used extensively in computer games and now also in other more sociable Web 2.0 systems such as Linden Lab's *Second Life*, where people can participate in an artificial collaborative virtual world and also create new artifacts and thus extend the world. This raises an interesting point about the extension of maps and positioning systems from the real world into virtual worlds. Linden Labs has spawned a real monetary economy (starting with a virtual one based on Linden Dollars) where participants trade their artifacts. Maps of virtual worlds may become interesting commercially (at least to the gaming aficionados). Virtual reality maps of the real world are also growing, which are used for "serious gaming" (e.g., scenario, mission, and contingency planning) and by planners and architects to showcase future built environments to prospective stakeholders.

8.5 Geotagging, Geoindexing, and Searching

8.5.1 Geotagging

In the context of the WWW and its information protocols, a tag is a reference embedded within the script that accompanies each Web page. It might become a visible entity such as a displayed image or remain invisible and form part of a logical address to something important. In the context of the Geoweb, a geotag is a reference to a geographical entity, for example a point of interest (PoI), and is associated with a pair of geographical coordinates. Automatic geotagging can be achieved with a GPS receiver that would perhaps store a waypoint that could then be converted into a geotag and included on a Web map. Web 2.0 mapping actively encourages people to add their own tags and enhance the usefulness of the content (particularly if shared). An example of a geotagged mashup would be to place geotags on a Web 2.0 map or aerial image in the place where holiday photographs were taken. The actual photographs might be stored within a Web 2.0 photograph sharing site (e.g., Flickr) but the URL for each photo along with other information such as time, subject, and direction would be associated with the tag.

8.5.2 Geoindexing and Searching

Using a less structured approach, WWW content may be automatically searched with the aim of recognizing geographical information (and indeed other information that is in a structured format). For example, a simple postal code or a more complex set of latitude and longitude coordinates could be extracted if they were always associated with addresses. So it is possible to take general-purpose textual content (that is scripted behind each Web page) and set an indexing engine to "crawl" through the pages and extract the georeferences, and then to build them into a database that can be then rapidly interrogated by various location-based services. This bottom-up approach is less predictable but can greatly improve the access to random information. For example, along with postal codes or coordinates, it would be possible to extract times (perhaps opening times) so that the database would hold information of all restaurants and fuel stations along with their locations and opening hours, which would be useful if a service were asking for all facilities within a 10-minute drive that are open. Titmuss et al. [21] describe the way to preindex location information so users (who know their location) rapidly get useful short lists of relevant information in response to their inquiries.

The most effective way, however, of dealing with structured geoinformation (or any other information space where there is a clear, rational approach and hierarchy) is to adopt a highly standardized approach. There are many standards and standards bodies in mapping: we shall now look at some of the more important to the Geoweb. Figure 8.3 shows how the Web and its protocols have been changing and how the future is likely to develop with reference to maps and GIS. (See Chapter 9 for explanations of Semantic Web that includes the use of RDF and OWL, and the technique of simultaneous location and mapping or SLAM.)

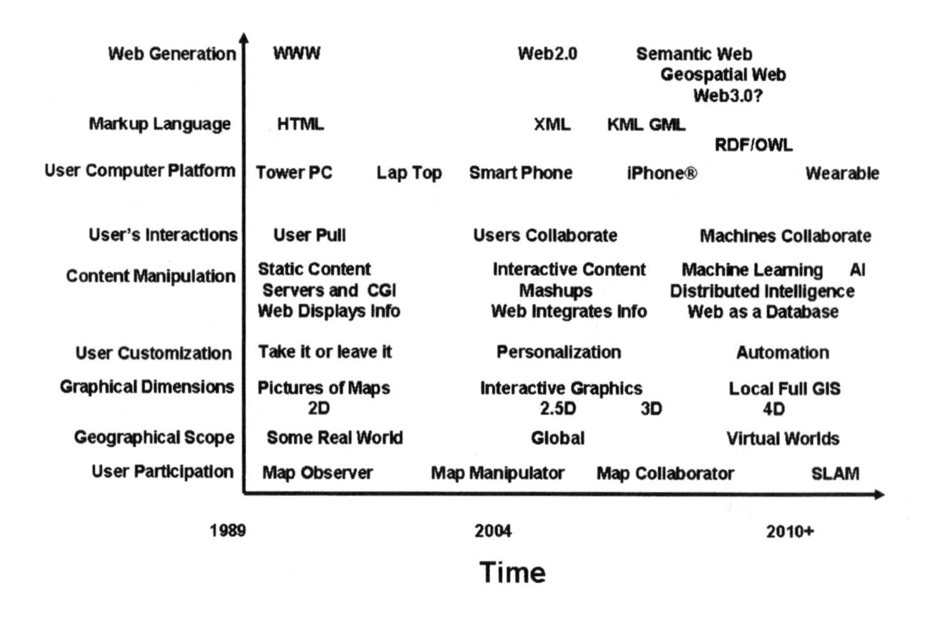

Figure 8.3 The Web and its evolution with respect to maps and GIS.

8.6 Standards

8.6.1 Markup Languages

Before looking at mapping, it is necessary to discuss aspects the World Wide Web and the way it is developing. When the Web emerged it was founded on several standards, two of which were the brainchild of Sir Tim Berners-Lee. Like all Internet applications, by definition the use of Transfer Control Protocol over Internet Protocol (TCP/IP) was needed to move packets of data from computer to computer reliably. To transfer files File Transfer Protocol (FTP) is used (and at its heart, each page and media entity of the Web is a file). The inspiration was to invent the Hyper-Text Transfer Protocol (HTTP) so that users could automatically fetch new pages (i.e., files) from servers where they are stored by "clicking" on URLs displayed on the browser application. Behind the scenes, in the script for the page the URL is spelled out in detail as the Internet address for the file. The second invention was the language for the scripts known as Hyper-Text Markup Language (HTML). HTML was extended into a powerful language capable of displaying information and allowing easy interaction with users via simple mouse controls. The main problem though was that an HTML script "knows nothing" about its own content other than how to display it. For example, it showed that a line of text was in italics but not that the text in question was a postal code, latitude, or longitude. If the HTML behind a Web page was about geographic information, the browser, server, or any other computer really could not tell. What was needed was something like HTML that could tag content to show what it meant (in human terms) so that computers could do something specific with it (in computer terms).

The next leap forward (and at the heart of Web 2.0) was the standardization of the Extensible Markup Language (XML). XML coexists with HTML so that browsers can interpret both (i.e., HTML was extended in its scope).

XML is a framework to define other markup languages that are domain-specific. XML is not about geography (or anything else) but it can be used to specify geographical things like a road, a point of interest, or a set of coordinates that make up a shape in a way that is meaningful to both humans and machines (i.e., the script can be read as it is written in plain text and parsed [or encoded on the fly] by machine software into a database). XML is used to define meaningful classes of "objects" and then to set real-world instances of the objects to actual values. (This approach draws heavily on the methodology of object-oriented computer programming, where abstractions are made about the raw data used by the machine, in order to make them more meaningful to humans).

Many new markup languages have emerged and this movement in computer science is gradually becoming the way most new computer information is being structured. For example, XML is at the heart of Microsoft's .Net initiative and is the way applications encode information. The next stage of the Web revolution, which is the main focus of Berners-Lee's activities in the World Wide Web, concerns the Semantic Web. The World Wide Web Consortium (W3C) is the main standards body for the Web and it is working on the necessary framework so that all fields of information can be more meaningful and accessible to both humans and machines and thus advance the science of machine learning and artificial intelligence (or AI). Chapter 9 will discuss its implication for Whereness.

Some areas of knowledge are ahead of others in the creation of markup languages but we are fortunate that the geographical sciences are an early adopter! Two geographical languages will now be described.

Keyhole Markup Language (KML) is the standard that is used by Google and is relevant to anyone using its API. It includes definitions of latitudes, longitudes, headings, ranges and so on, and is aimed at geographic visualizations.

Geographic Markup Language (GML) is more general and is a standard defined by the Open Geographical Consortium (OGC) that describes geographical features. It can be used for several purposes including computer modeling (i.e, creating software models of either an existing geography or a virtual geography to solve some problem) or for the interchange of information between proprietary GIS databases. The OGC is extending its standards into new areas, for example, creating a sensor modeling language for sensing systems.

8.7 Open Mapping

In this section we consider the role of openness in mapping, as opposed to a conventional commercial and proprietary approach where how and what is done are trade secrets. Openness assumes that although some things may be patented and require licenses, these will be made available so that a standardized approach may be adopted across an industry. Other things will be freely available but legally protected to prevent their being wrapped up within proprietary products.

If Web 2.0 is changing fundamentally the way in which digital maps are used by people, the advent of open mapping is likely to change the future of the maps themselves. As GPS-enabled consumer equipment gets cheaper and more intelligent, the ability to collect geographical information will grow, especially if the function becomes entirely automatic. In Section 8.5.1 we have seen how traffic congestion can be detected by navigation units; a logical extension is to use the GPS track-logs to make maps and then to share them.

Traditional maps are made by ground surveying, aerial photography (either from aircraft of more recently from space), using lidar (light, detection, and ranging), which is the optical equivalent of radar, and by professional satellite positioning equipment. Small numbers of highly skilled professionals use expensive equipment to produce high-quality (expensive) maps. In contrast, amateur mapmakers use very cheap and less accurate equipment but have the advantage of using "crowdsourcing." This is the term used to describe an approach where motivated amateurs are self-organized to cooperate to achieve a goal, the result of which is covered usually by a legal framework that allows free use but not commercialization. By combining the logs for a number of tracks, the inherent inaccuracies of consumer GPS can be averaged, producing very useful maps.

The terminology "the wisdom of crowds" has been adopted to describe the sentiment which supposes that by using a large number of independent people the average will be useful. There is a counterargument that experts are still needed to validate and verify information, especially where there may be malicious intent. Geohacking is in its infancy but may well become a nuisance, but more serious manipulation of maps for criminal, political, or military ends is also a possibility.

8.7.1 OpenStreetMap (OSM)

OSM [22] is building up an impressive range of free maps created by amateurs. The organization both hosts the content and provides the software framework needed to create the maps from track-logs and feature attribution.

8.7.2 Publicly Funded Mapmaking

In the United States, maps from the U.S. Geographical Survey (USGS), which is an arm of government paid for by general taxation, are given to citizens freely. In contrast, in many other countries (e.g., the U.K.) government maps are sold commercially. It will be interesting to see how the different funding models affect the quality of mapping, particularly for countries such as the United States which has a vast territorial extent. It may be that open mapping by citizens compliments the public approach whereas the more commercial operations would be disrupted. Open mapping is a genuine disruptive technology that could curtail at least some traditional commercial map operations.

8.7.3 Infrastructure for Spatial Information in the European Community (INSPIRE)

A different approach to openness is the top-down approach being adopted by the EU in 2007, to address the fragmented approach to maps across the Community. Its aim is standardization and interoperability and it may become a very significant agent for change.

8.7.4 Agencies to Validate and Verify

It is not clear yet exactly how the mapping and ICT industry, and governments will be dealing with the need to control open mapping. It seems reasonable to suppose that industry groups will to an extent self-regulate but given the strategic importance of reliable mapping it is likely that governments will see the need to be active (as we saw in the section above). Although open mapping may be free, there may be other versions of the same maps that have been checked where a fee is payable (to cover the costs of checking). Organizations with trusted brands will be prime candidates to perform this function which may become the main role for today's traditional mapmakers.

These organizations may actually start to promote open mapping and crowdsourcing as a cheap way to get an area mapped (and remapped regularly, since it is the changes to maps—their maintenance—that is expensive). The quid pro quo is that the free maps will be hosted and available for all in a verified and validated form for mission-critical applications.

8.8 4D Maps and the Temporal Dimension

Dynamic information concerns things that are changing rapidly with respect to time. Adding dynamic geospatial information to static maps was discussed in Section 4.2. A spatial map may have three dimensions but the addition of time adds a further dimension. For example, adding to each road link on the map its real-time road congestion (as an impedance factor) allows a computer to calculate not just the shortest route but the quickest route. Another transport example would be the progress of a specific vehicle along a route that could be displayed as a moving image overlay to the static map which remains stationary on a screen. A good example of a dynamic map is *Traffic England* [2], where the U.K. Highways Agency links the Web map of the U.K.'s main routes to the highway traffic sensors and the messages displayed on the roadside variable message signs. The map changes in real time (minute by minute) as the various events unfold. The M25 motorway around London is one of the most congested roads and its problems at busy times (i.e., most of the working day) result in large amounts of dynamic information.

8.8.1 Floating Car Data

To map the dynamic state of vehicle congestion on roads it is necessary to sense the traffic flow. Many types of stationary traffic sensors are used, including inductive loops beneath the road surface, optical cameras, and infrared detectors mounted above the road, but the most efficient option is to use the vehicles themselves. If fitted with a GPS unit or other positioning system, the vehicle becomes a sensor that can report dynamically in real time using mobile communications (e.g., the GPRS service of GSM). The data thus collected is known as "floating car data" or "probe data," and in early experiments it was found useful to use fleets of taxis that randomly "float" around a city. An alternative, more controlled situation, can use buses to probe specific routes regularly.

It is generally considered that if just a few percent of moving vehicles are probing, a useful statistic is collected that can be fed back to drivers (via their navigation units) to help them avoid traffic jams and thus improve the overall traffic flow efficiency.

8.8.2 Calendars, Diaries, and Logistics

The movement of people can also be managed by a 4D map that can be considered as part of an electronic diary or calendar. Already there is much static groupware (i.e., collaborative software applications), for example the personal calendar in Microsoft Outlook. If these are combined with a personal positioning system, perhaps based on wire communication access (using WiFi hotspot or cellular radio), and a GIS, maps can be made on demand to show where people are located (with varying degrees of likelihood as the a priori nature of diary entries is confirmed by physical measurements).

The logistics of physical objects of high significance can be treated in much the same ways as for people, except rather than a calendar database there will be custom enterprise software that performs the event processing wrapped up in solutions associated with supply chain management such as stock control and warehousing.

8.8.3 Event Processing

Overall, however, there is a need to converge all these locations and event functions into an overall system of Whereness, so that the common functions of 4D mapping and event processing can benefit from economies of scale and conform to an open standard.

An event engine can be added to systems using GIS, for example, a traffic management system or a groupware calendar/whereabouts system. Both need to generate events in the form of messages or software processes, in response to specific temporal-spatial combinations. For example, a dispatcher may wish to know when a specific truck has reached a way point to predict if a delivery will be late or on time. A manager may wish to see if the planned meeting is viable given the progress the attendees are making with their various routes and personal schedules. A police department may wish to be alerted if an unusual traffic pattern of vehicles or people is being detected that may be indicating a forthcoming crime that could be as serious as an act of terrorism.

The importance of 4D maps in creating value is clearly very high but it does create serious technical challenges, particularly the issues of scalability, latency, and availability.

The scale issue concerns the burden of processing (and investments to provide it) required to perform regular 4D calculations to check the status of each event instance being monitored. In the extreme there could be several event types being monitored for each driving vehicle, active person, or important object. Latency (or the delay in processing or communications) if too great can lead to time critical deadlines being missed or confusion about whether an event has happened or not. Availability of networks, positioning signals, or mapped areas leads to patchy services and unreliability. These challenges are to an extent linked but all contribute to the overall quality of service (QoS).

8.8.4 Carrier Scale Whereness

A strong incentive to use a converged Whereness system is that QoS can be managed more effectively and efficiently. Sometimes the term "carrier scale" is used to describe the robust nature of the trunk telecommunications networks that underpin international telephony and the Internet. Whereness operated on a similar scale by similar organizations should lead to a similar QoS.

8.8.5 Time Calculations

If we wish to include time in the maps, it is likely that time will be represented by a period and not by an instant, so we need four values for the upper and lower bounds of the start and end of the period in question. This is needed to help deal with queries such as, "When the planned roadworks are in progress, will the truck in question be delayed?"

The road would be defined as the ribbonlike area defined by the nodes representing its geometry and the truck as a position probability (probably an ellipse, since positioning errors are often better in one direction), but both road and truck would be qualified by time bounds. For the truck the timing may be based on its GPS tracklog.

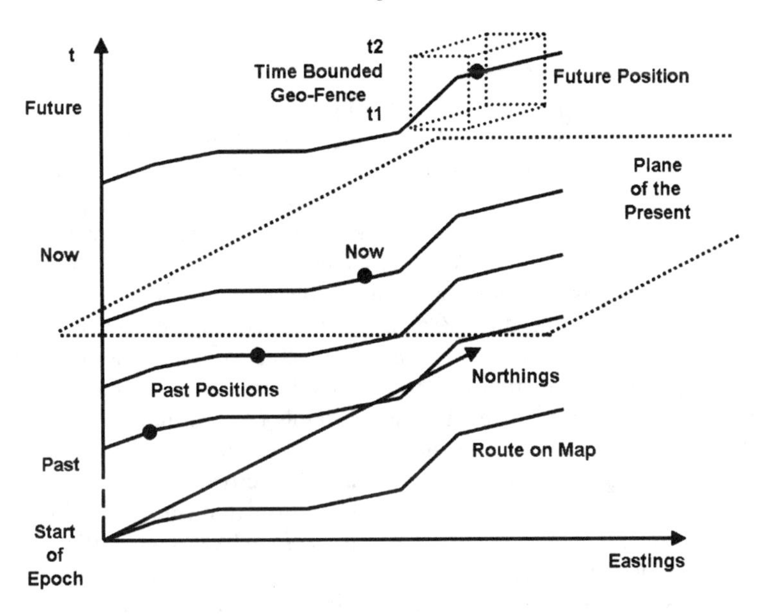

Figure 8.4 Map with the third dimension of time on the vertical axis.

Figure 8.4 shows the information as a 3D graph with time on the vertical axis. The route on a map is assumed to be on a horizontal plane and time can be considered to be horizontal planes with the plane of the present shown with dashed lines. The beginning of the epoch is shown at the origin of the graph and the progress of the truck shown on the route at different times. Sometime in the future a geofence is drawn on the route to show where the roadworks will be and the height extended to form a rectangular box (shown in dashed lines). The event-processing software would be calculating the probability of the vehicle actually passing through the box .

This diagram is helped by the fact we are ignoring one spatial dimension (i.e., altitude), which is assumed to be constant. Also, the route is constant and the only changing entities are the truck's positions on the map expressed as north-south or east-west coordinates on the map (i.e., Northings and Eastings, respectively).

The event engine would have huge numbers of similar boxes defined and processes that trigger at set intervals to check the boundaries. Trigger rates will depend on QoS and application requirements. For a fast-moving vehicle a few minute's delay can represent a significant distance.

Chapter 9 discusses the longer term aspects of mapping that are current in the research phase: first, the use of the Semantic Web to create an ontology for Whereness as part of a standardization process, and second, being able to simultaneously locate position on an existing map, correct it if found to be in error, or extend it if the measured position is unmapped.

8.9 Summary

In this chapter we have seen how traditional maps have changed from pictorial images to digital informational spaces that can be manipulated in GIS software that uses a vector representation of space. This has enabled the maps to move from flat 2D graphics and become true 3D representations that can be manipulated on a computer screen to provide a much more meaningful user interface. We have seen the fantastic strides that recent Web 2.0 techniques and organizations like Google have made in bringing digital maps to ordinary people who can use a global resource with minimal need to learn any skills or pay any charges. Mobile devices such as the iPhone mean the maps can be used on the move and customization means the information will be what the user wants, delivered when it is wanted and in the desired format. The map metaphor is becoming the portal into the world's information and is a disruptive technology as all manner of other Web 2.0 services gain a geographical and positional perspective.

Service providers are exposing their underlying information via APIs and thus allow third parties to mashup data to create new services. AJAX is a key Web 2.0 technique that empowers the user's browser so that ordinary people can have a powerful GIS-like system without buying or installing any new software.

Maps are also becoming dynamic and adding the fourth dimension of time so that real-time changes and the needs of moving users can be met. New services based on event management are now possible. Further dimensions extend mapping into virtual worlds and augmented reality where the virtual worlds can overlay the real world.

Users can become map content creators and soon will be able to observe themselves on the map as well as use it. Traditional mapmakers are no longer the only organizations with high-quality digital maps.

The standards activities that underpin the Web are leading to new mapping markup languages and the Semantic Web is beginning to provide a framework to bring all of Whereness together.

References

[1] O'Reilly, T., "What is Web2.0," http://www.oreillynet.com/pub/a/oreilly/tim/news/2005/09/30/what-is-web-20.html, Dec. 2007.

[2] UK Government's Highways Agency, *Traffic England*, http://www.highways.gov.uk/traffic/traffic.aspx, Dec. 2007.

[3] *Google Maps*, http://maps.google.com/, Jan. 2008.

[4] *Google Earth*, http://earth.google.com/, Jan. 2008.

[5] Blakely, R., "Microsoft Challenges Google's Dominance with $50m Deal to Buy Multimap Website," *The Times*, 13 Dec. 2007 pp. 51.

[6] *Multimap*, http://www.multimap.com/, Jan. 2008.

[7] *TomTom*, http://www.tomtom.com/. Jan. 2008.

[8] Teleatlas, http://www.teleatlas.com/index.htm, Jan. 2008.

[9] *NAVTEQ*, http://www.navteq.com/, Jan. 2008.

[10] *Mapquest*, http://www.mapquest.com/, Jan. 2008.

[11] *uLocate*, http://www.ulocate.com/, Jan. 2008.

[12] *aboutmyplace*, http://www.aboutmyplace.co.uk/, Jan. 2008.

[13] *ViaMichelin*, www.viamichelin.com/, Jan. 2008.

[14] *meetro*, http://www.meetro.com/, Jan. 2008.

[15] van Essen, R., *Maps Get Real: Digital Maps Evolving from Mathematical Line Graphs to Virtual Reality Models*, 3D geoinfo07,Delft NL., http://www.3d-geoinfo-07.nl/documents/program.pdf, Jan. 2008.

[16] Oosterom, P., van Zlatanova, S., Penninga, F., and Fendel, E. (eds.) *Advances in 3D Geoinformation Systems, Lecture Notes in Geoinformation and Cartography*, Berlin, Heildlberg, Springer-Verlag, 2007.

[17] Ballard, P.,*Ajax in 10 Minutes*, Indianapolis, IN; SAMS Publishing, 2006.

[18] Erle, S., Gibson, R., Walsh, J., *Mapping Hacks*, Sebastopol, CA; O'Reilly Media, 2005.

[19] The Economist, "The world on your desktop," *The Economist Technology Quarterly*, Vol. 384, No. 8545, pp. 16-18.

[20] Gibson, R., Erle, S., *Google Maps Hacks*, Sebastopol, CA; O'Reilly Media, 2006.

[21] Moore, R. P., Titmus, R. J., Allen, K. M., Lebre, C.A.M., Telecommunications Apparatus and Method, US Patent 6397040, Aug. 2000

[22] *OpenStreetMap*, http://openstreetmap.org/, Jan. 2008.

Chapter 9

Whereness and the Future

This concluding chapter is about the longer term aspects of Whereness. The first part will continue with the technologies that were discussed in Chapters 6, 7, and 8, and will concentrate on more advanced and speculative topics which are still in the research phase. Although they may not all become commercial successes, they have the potential to be very significant.

The second part will focus on applications, users, and businesses in the 2010s: first, a longer term vision discussing devices and services, followed by a to-do list for business so that the Whereness vision may be, hopefully, followed by action.

9.1 The Future of Wireless and Whereness

Wireless science is still being researched and there are many new developments to come. Devices will increase in the upper frequency of operation so that new mobile bands will become available in the region of 60 GHz and beyond. Terahertz radio is in its infancy but is finding applications in radiography. New techniques based on the application of ubiquitous computing are developing [1] and also new approaches to the management of spectrum. Antennae are becoming smart and many military techniques will become cheap enough to use for consumers and for business applications.

9.1.1 Atmospheric Absorption Bands

In the region of 57 to 65GHz, radio signals resonate with oxygen molecules in the atmosphere and are severely attenuated. In normal radio practice, environmental attenuation is a major problem but in this case it can be an advantage, particularly for Whereness systems in the future. A new coverage paradigm could be implemented in this band, which would result in very dense localized hotspots, each with massive bandwidth capabilities.

Conventionally, a major spectral problem for a mobile operator is frequency reuse. This is the problem that signal strength does not fall in level enough with distance once past the edge of cells. Before the channel can be reused, it must be low enough in level to prevent cochannel interference so patterns of reuse are planned so that sometimes several cells are needed in between (using other channels) before the original one can be reused. With atmospheric absorption, the reuse distance is very much less so channels can be reused several times within an area of a square kilometer. This leads to a massive increase in the information that can be communicated.

The main reason for holding up the adoption of this paradigm is the devices, which are currently exotic and expensive. As semiconductor science advances it is likely that new cheap devices may emerge, perhaps using nanotechnology.

From a Whereness perspective the prospect of a high density of very wideband transmitters is ideal. In the built environment, these hotspots may be built into the fabric of the buildings together with new LED lighting. Truly massive bandwidths for mobile communications would result with potential also for sensing and positioning.

9.1.2 Chaos and Convergence

The future of wireless looks chaotic as more new standards emerge and commercial rivalries continue as wireless Internet access and cellular radio standards overlap in capability. For example, GSM operators are deploying domestic GPRS picocells to compete with open-WiFi. All these issues are beyond the scope of this book but the recent "Mobility and Convergence" issue of the *BT Technology Journal* gives a good overview of digital radio preceded by Hodgkinson's useful introduction to digital wireless [2].

New spectrum will be available (at a price) for new approaches such as WiMax and similar higher powered and longer range wireless LAN technology. All of these activities will add further opportunities to use wireless signals for positioning and for Whereness applications to exploit the increased wireless coverage. The need for convergence will increase because users do not generally want to deal with the technicalities of manual network selection and will just want services to configure automatically, regardless of which sort of mobile wireless access network is used and who is providing any applications. Considerable work is still needed to make device and service simplicity, via automation, a reality.

Portable equipment will become increasingly multibanded with their core wireless components being affected by the recent software defined radio movement. Most digital radios use similar processing blocks to perform functions that include filtering signals, modulation and demodulation, generating signals, and controlling smart antennae.

Traditionally, each of these intensive digital processing tasks was performed in dedicated silicon subsystems or application specific integrated circuit (ASIC). With software defined radio, however, common digital signal processors (DSPs) may be used within a wireless transceiver and configured by software, so that the common blocks can be assigned to tasks according to the demands of each type of radio system. Instead of a computing platform of today with many separate subsystems (with one for each standard), the future will consist of a set of more general-purpose processing blocks that can be configured and reconfigured depending on the mix of standards and protocols that are needed.

The requirements of positioning will be added to the list of radio processing tasks that software defined radio will be performing. These include measuring signal flight times (for lateration), processing antennae arrays to steer beams (for angulation) and collecting multiband signal strengths (for proximity estimates). Although there will be radio hardware in the digital transceiver of the future, it will be more versatile so that when a new radio standard is needed or a new processing function required (perhaps for a Whereness service), the platform will not need to be modified physically since the changes will be all in the configuration of the DSP elements performed by software drivers. It is unlikely that true software radio will ever be common (i.e., all radio processing done in software) but the software defined radio, with the DSP array approach, is both economic and pragmatic and a very useful way to converge standards and functions.

9.1.3 Cognitive Radio

Cognitive radio is a new paradigm for the self-organization of radio spectrum, radio coverage, and radio systems. Whereness could have a profound effect on the more advanced configurations.

Currently, radio coverage and spectrum allocation are organized centrally. Regulatory bodies, who many years ago allocated single channels to individual organizations, now tend to allocate blocks of spectrum to trunked operators who then share out individual channels on demand. A centralized network management system is used to control the trunking and radio channel allocation. Full cognitive radio takes the approach that each radio itself should make the decisions about the allocation of resources, based on local knowledge of the actual state of spectrum occupancy. This is a decentralized approach and involves the radio nodes having the capabilities to sense, scan, and operate (peer-to-peer) with other local nodes, perhaps avoiding many of the centralized control functions present (e.g., in a cellular radio system).

Supposing Whereness was widespread and by a variety of means, some of which may not be wireless, mobile nodes know exactly their location and the location of other nodes. Supposing also each node has a detailed map that includes a terrain model. It would then be possible for any node to establish a wireless link to any other viable node with optimal power, direction, band, and modulation. No energy would be radiated that was unnecessary and overall co-channel interference levels would be reduced, greatly improving frequency reuse and spectral efficiency.

Currently, each trunked radio link is used with an estimated power level that is usually adjusted by the system to maintain an adequate bit error rate, but with this ad hoc proposal, every link would be optimized. There really is no practical limit to the amount of information that may be passed by radio but the apparent spectrum congestion is mostly due to planning inefficiencies.

In the past, the need for terrain models in radio planning has been a once only exercise required when the radio network is built (and occasionally afterwards when new base stations are added or moved). Cognitive radio with embedded Whereness implies that the radio planning tools are embedded within the dynamic operation of the system and are used every time any transmission is attempted. Every successful radio transmission will therefore confirm the accuracy of the mapping and positioning and conversely, every failure will suggest to the system that the terrain model or the positioning estimates are incorrect. Cognitive radio could therefore also be a useful source of positioning information.

Ad hoc networks are likely to become very common, and they will use aspects of cognitive radio (if not the full capabilities). For example, it seems likely that personal music players will soon be able to share media by wireless informally. Currently, the topology of an ad hoc network would be determined by a trial and error approach. A mobile node would try to communicate and find its neighbors and they would find theirs and so on. With positioning factored in, all local nodes would potentially know the locations and status of every node so there would be less wasteful trial and error. With motion models factored in, a device might well hold off communicating until it knew that another party would be in range (given the current knowledge of the terrain, and the predicted motion of both parties).

In essence, what we are suggesting is that radio systems become "users" of Whereness as well as providers of position.

9.2 Sensing Futures and Whereness

There are likely to be many advances in sensing technology based on current research in biotechnology and nanotechnology, some of which will be relevant to positioning and many of which will mimic nature. Also, so far we have only considered electronic ICT but there is a growing trend that concerns the use of nonelectronic information systems where physical effects other than electricity and radio are used. One fruitful area concerns optical systems where photonics is used to identify and process information. For example, "tagents" are substances that are used to track and trace objects as an alternative to conventional RFID technology. They are mixtures of dyelike chemicals where the mix corresponds to a code, rather like the stripes of a bar code except all the fragments can be detected at the same time by an optical infrared spectrometer. The presence of each dye fragment corresponds to a line within the spectrum.

Another barcode-like approach to sensing position involves the detection of patters by cameras. Animals recognize objects by shape and form and a simplified approach is to use optical barcode-like shapes that are very cheap to place around an environment. Harle and Hopper [3] describe an experimental in-building positioning system based on this novel idea that uses clusters of visible markers attached to walls and objects.

Olfactory tagents are not a current technology but nature uses this method for many purposes to help many animal species to find the position of food, mates, and rivals, and to follow trails to food supplies. Once effective artificial chemical sensors (i.e., artificial noses and taste buds) are available, these systems may become useful. A promising approach might be to adapt the technology used to detect explosives in security scanning.

Nature also appears to have very effective processing systems. The eyes of flying insects have neural networks that appear to be able to detect and process the real-time geometry of a scene to facilitate expert feats of 3D navigation. Sensor fusion must be taking place since some of these insects also have gyros and olfactory sensors. This area of artificial intelligence, when understood, may have huge potential for Whereness. At present, multisensor fusion is a heavyweight processing job but clearly nature can do better.

In addition to the mimicking of nature, another approach would be to extend it. Humans have five normal senses, and these could be extended artificially by the use of additional sensors carried about or within the person. A sense of direction is a somewhat vague term, but it would seem useful to integrate a full personal guidance system with the body. For example, the interface could be mostly hidden using touch technology and commands issued by natural voice. A sense of a co-located group could be made explicit in much the same way. Some animals flock together or run or swim in groups, and as social networking technology advances the sense of a local group (based on the position and context of all its members) may be desirable. Although nature does not use radio, there is no reason not to use it to extend our own sensing by communicating and reacting peer-to-peer with colleagues, friends, and sentient objects. In the future, ubiquitous or pervasive computing research suggests myriads of internetworked objects in many spaces.

One application that may emerge for security reasons would be the personal black box sensing recorder that would be rather like a human version of the flight recorder found in aircraft. It seems likely that vehicles would also benefit from such a device to resolve disputes when accidents or regulation infringements take place. The personal (or vehicular) universal sensor would collect position and orientation information, acceleration and rotation, images, sounds, light levels, the presence of noxious substances, radiation, and anything that is harmful and useful as evidence to a court of law.

This sensing platform would be ideal to deliver everything of an autonomous nature concerning Whereness. In particular, it would offer the potential to greatly enhance the mapping of all environments.

9.3 Intelligence and Whereness

In the concluding paragraph of Chapter 7 the problem of the fragmented nature of the information resources needed for Whereness was highlighted. All the necessary individual Whereness functions (i.e., converging information through data fusion, following previously mapped information, and mapping information collaboratively), still require an overarching framework to the information itself. This leads us into the field of artificial intelligence and the need to create a standardized Whereness framework or ontology.

9.3.1 Taxonomies and Ontologies

When an area of knowledge is being considered, the first stage of the process involves the creation of taxonomies and ontologies. Taxonomy[1] is an informational hierarchy concerning the terminology. A positioning taxonomy has already been published and was discussed in Chapter 2. An ontology, which is a term used in philosophy concerning the meaning of being, has been adopted by computer scientists, and in this context is useful as a standardization process. Once adopted, there are then automatic computer software tools to greatly facilitate the ability to use the information in computer applications without the need for specialized programming efforts. In essence, basic knowledge is abstracted into a higher layered framework that is then more generally applicable. The great advantage will be the automation of the process of finding, combining and using information (i.e., without the need for new software to be written). Whereness seems an ideal candidate because of the fragmented nature of the component parts that encompass the domain.

Some knowledge domains are ahead of others in using the Semantic Web but fortunately the area of geographical information is at the forefront with efforts coordinated by organizations such as the OGC and standards such as GML (see Section 8.6.1). It is reasonable to suggest that real-time positioning could be part of that domain or perhaps both part of a larger domain that includes everything concerning time, space, radio, and sensing. In these sections the formal stages for the creation of a Whereness ontology are explored.

9.3.2 The Semantic Web

The most promising relevant methodology in computer science concerns the Semantic Web, a movement influenced by the Web's creator, Sir Tim Berners-Lee [4] who leads W3C [5], the body that seeks to standardize the future of the Web. In simple terms the aim is to turn the Web from a mainly human readable resource into a database that is meaningful to both humans and machines. It has been suggested, therefore, that each domain of knowledge be analyzed by domain

---

[1] A taxonomy for ubiquitous positioning is included in the Epilogue.

experts (in cooperation with Web experts) to create a set of standards to define everything that is relevant according to the Semantic Web methodology. The techniques are new and the standards still emerging but any new domain of knowledge should be seriously considering its impact.

The benefits are about the automation of converging and integrating diverse information spaces, which should greatly reduce the costs of the ICT systems and lead to an open and nonproprietary approach that should in turn lead to greater opportunities. Just as the original Web led to an explosion of information sources for people, so should the Semantic Web do the same for machines (acting on the behalf of people).

9.3.3 Why an Ontology?

A set of tests have been published [6] to help when considering if a potential domain fits the need for an ontology. There are five things to consider:

- Sharing a common understanding;
- Reuse of knowledge;
- Making domain assumptions explicit;
- To separate domain knowledge for operational knowledge;
- To analyze domain knowledge.

What would the domain be? In our case (potentially) it would be Whereness. What would it be used for? Finding positions in real time and sharing the information. Who would use the information? People and things that are moving (or are interested in things that are moving) either personally or via supporting service providers. The next stage is to answer some competency questions.

9.3.4 Ontology Competency Questions

It is easy to generate questions but answers will be coming from ongoing research and standards work over the next few years. The following questions are likely to be the most important:

- Is there enough information to determine position? Although the answer may be a theoretical yes, practical considerations may be making it difficult because of system or commercial issues. The Semantic Web approach may well ease these problems since an open standard is being used.
- Given any physical measurements that exist and that are accessible, can a location be found on an available map?
- If there is no map, is there now a reason to consider creating one to be used in the future?

- If sensor readings and tracklogs are available, can mapping commence?
- If there is a map available, is a more standardized approach to access useful?
- If the map is available, is it in a form that can be used to translate the physical information into the symbolic so that sense can be made of the information by people or machines?
- Given a reading of position, can its accuracy be increased by reference to other measurements taken by systems not yet available?
- Can any map being used be improved or extended?
- If an inconsistency is detected, can steps be taken to correct it?
- Are there existing ontologies that can be included? For example, from the domains of GIS (with GML) and sensing (or SML)?
- Should information spaces concerning identity such as friend of a friend (FOAF) be included?
- Could geographical information such as postcodes or geotags be included?
- There seems to be a lack of standardization concerning general radio issues such as location and frequency of transmissions (although some progress is being made with WiFi hot-spots), so is a radio markup language viable?
- Would it be useful to expose the various cellular radios positioning systems information in a standardized way?

9.3.5 What Would the Ontology Look Like?

Fortunately, considerable research effort has been given to help in the creation of ontologies, and a Web Ontology Language (OWL) has been created. Lacy [7] describes the process of using this tool and all the other supporting frameworks. In common with much computer science, it is a layered approach where one set of concepts supports another at a higher layer of abstraction. OWL is at the top.

OWL is used to describe the domain semantics, and several domain building blocks are needed using concepts borrowed from object-oriented programming. "Classes" define important entities and are defined first. These are probably explicit within the taxonomy. For a Whereness ontology some top-level classes could be:

- <radio base stations>[2]
- <maps>
- <mobile stations>
- <tag readers>

---

[2] <...> notation is used to show that the text between the brackets is not part of the narrative but is a parameter, delimited by brackets.

Each of the classes could have subclasses and those subclasses could have subclasses, and so on in a treelike hierarchy. Thus <radio base station> could have a subclasses <long range> and <hotspot>. The domain experts would decide on the definitions and these would be recorded in the standard.

The second building block is property. Within a class or subclass, an entity may have properties; for example, a <hotspot> may have a <hotspot standard> since it could be a WiFi hotspot or a Bluetooth hotspot (or in the future some other as yet unknown standard). The advantage of this approach is that it lends itself to modification and extension and is thus future-proof.

When an actual situation is being defined, the building block of "individuals" or instances is needed. For example, a specific hotspot will need its <type> to actually have a value (i.e., <WiFi>). A specific base station would have a mapped location that would take the value of a specific latitude and longitude.

Following the OWL ontological primitives there needs to be a way to deal with the resources. For example, the definition of a base station will actually be located as a Web resource, all of which have Uniform Resource Identifiers (URIs). The familiar Web page Uniform Resource Locator (URL) is a type of URI. To make sure everything described is unique and not ambiguous, the concept of "namespaces" is used. When a URL of any sort is examined in detail, it can be seen that the format /something/something-else/ … is used so that although rather long and tedious, the actual URL is absolutely unique since the namespace is always extensible (and thus can be made unique).

A method is needed to create real low-level relationships so that values can be assigned. The syntax is described in XML, which does not describe anything per se but rather is a framework used to define domain-specific markup languages. For example, if GML and SML were viewed as text on a page, their formats would be similar but the specific terms would be unique. Stahl and Haupert [8] describe how XML-encoded indoor building models can be useful for positioning systems. If the actual data is viewed, it will appear to be in the markup scripting form that is standardized as the Extensible Markup Language (XML). When viewed, it can be seen to be similar in format to the earlier HTML of the early Web but is much more versatile. (A good way to see these markup languages is to view a Web page on a standard browser and on its toolbar select <view> then <source> .)

To be able to generate lists of data to be included within any instances of a Whereness ontology (which is really a distributed database) there needs to be a way to associate objects (i.e., instances of the classes or subclasses) with appropriate properties and their values. The Resource Description Framework (RDF) is used. OWL describes terms of the ontology and their hierarchies, but it is RDF that actually deals with the information (and uses the domain vocabulary that OWL was used to define). Each line of RDF is rather like one record in a database: lists of statements serializing the real-world information.

RDF is a powerful concept and it predated OWL and underpins much of the Semantic Web. Powers [9] describes the practical use of RDF. Three pieces of information (or "triples") are associated. First a subject is given, second a predicate (or property), and finally, a value. For example:

<my workplace> <has a country> <UK>
<my workplace> <has a postcode> <IP5 3RE>

These fields can be pointers to the information in the form of URIs to show where on the Web the information resides. Specific software then uses the URI to collect the data from the file pointed to by the URI.

The ontology is thus a set of standards that knits together information in standardized formats that can be located anywhere on the Web. It reads as a set of rather cryptic textual statements (like any other computer language), but within its structure are located all the information any machine could need to do the tasks associated with the ontology. Unlike a computer language, there are no functions or methods, just a description of information, where it is to be found, and what the actual instances of the information have as real-world values. The ontology could have pointers to resources where programs reside that can act on the data, but it is not an application that will do anything.

9.4 Mapping Futures

9.4.1 Simultaneous Location and Mapping

Simultaneous Location and Mapping (SLAM) is a technique for following and building maps at the same time and came from research work on mobile robots [10]. The technique is also being applied more generally to people who are carrying mobile positioning technology using the idea of "people as robots," where it is assumed that a person is guided by following instructions issued by the guidance computer. It is assumed that a digital vector map is being followed.

The first stage of SLAM is when positioning measurements are taken for map matching, and then positioning errors are also recorded. If new features are encountered that are not mapped, for example a new corner in an open plan office, then the feature is assigned an identity which will have a probability that is initially assigned a low value. When the feature is encountered again on other journeys, then its probability is increased until a threshold is reached when it is actually added to the map. From the error logs the opposite situation can be determined and used to remove features that are either erroneous or temporary.

Map matching systems using dead reckoning and inertial navigation can have the problem of error accumulation, so SLAM also looks out for landmarks to correct the errors that can be detected by secondary positioning systems (e.g., passing a hotspot or an RFID system). Another technique is to use a salient map feature such as a sharp corner within a route. Lessons can be taken from the way people navigate, which is often by way of landmarks that may be detected by various sensors.

For example in their paper Schindler et al. [11] describe a very simple positioning SLAM-like system that uses a wearable device consisting of a modified set of spectacles. It detects doorways between rooms by shining infrared beams sideways and looking for the characteristic reflection with distance being measured by a simple accelerometer chip-based pedometer.

Mapped information must be self-consistent and topologically correct. If loops are mapped they should be made to close up. If a multiple set of tracks all slightly offset are detected, they can be condensed into a single track. The great strength of the technique is that it is a closed loop control system. If the post-processing stage performs a consistency edit, it will be confirmed or rejected when the map is next reused.

The main difference between a robot exploring and mapping a space and a human "robotics" mapmaker is that the former may move into uncharted areas just to explore whereas it is more likely that people will be following definite routes with a purpose in mind. Human mapping will tend to map where people are likely to go, whereas a robotic explorer may be sent on more random mapping mission in open spaces.

Spaces are sometimes processed as occupancy grids. The likelihood of being in a certain place (based on measurements) is calculated statistically for each grid square in a given area. Another technique is to convert free space into a non-regular graphical grid and use a Voronoi diagram. Obstacles are assigned as grid points anywhere they are detected and equidistant loci between the points are plotted as places where it is possible to be located.

9.5 A Long term Vision

In this book we have covered a large number of concepts, many of which are speculative and suggest a future computing and communications environment that is unfamiliar today. So what might the developed world be like, for example, in ten years' time?

In a decade of so we can expect basic computing devices to be faster, smaller, cheaper, and more common, in line with the general trend that is supported by Moore's law arguments. The processing and storage tasks that would be associated with quite expensive personal appliances such as vehicle navigation units and personal smart phones would be included in much more basic units, leading to the possibility that virtually every working person and school student would carry a personal positioning device that would receive a basic set of signals from several GNSS systems and have connectivity by both cellular radio and hotspots. These devices would be used to organize intelligent transport at a regional and national level where the environmental impact of most journeys is calculated and tolled.

Security would also be assured with some urban and built environments using the basic device as an access and security tag. LBSs would be offered by many service providers and retailers with mobile advertising taking the cream of the advertising revenues. Locative social networking would be common and most groups of friends and families would routinely know of the locations of members, at least at certain times. People will share media, maps, and locations in a viral manner. Devices will also act as personal collaborative mapping tools with anonymous track logs collected invisibly as part of the service. The people who today routinely use texting and cell phone voice services (which will be much less expensive in the future), will be using new Whereness applications. Their average inflation adjusted revenues per user (ARPU) will probably be constant as Whereness revenue fills the revenue gap.

More upmarket devices with truly ubiquitous positioning would be used by niche groups, including business travelers, professionals, workforces, gaming enthusiasts, and early adopters. People today who have a Blackberry or iPhone will have a device that allows it to know positions virtually everywhere at an accuracy of around 3m. The devices will include capabilities to use almost every radio signal and to sense a wide variety of phenomena. Most workplaces, campuses, and some homes will be fitted with smart building systems that will react with the more expensive devices.

People who choose or cannot afford any of these portable devices will still be able to experience a limited set of applications based on environmental sensing systems. When anyone travels by public transport, the smart ticketing systems, video surveillance, smart flooring, and other fixed sensing systems will monitor position and other contexts for the prevention of crime and terrorism. (Common crime may fall but will be offset by a more serious problem as Whereness technology is used illegally.)

The whole of the built environment will be mapped in great detail, and these maps will be freely available and automatically extended, updated, and maintained. Quality assured versions of the free maps will be available at a charge from traditional mapping organizations whose businesses will now include indoor mapping, 3D environmental modeling, dynamic mapping, and imaging. Many new startup companies will be extending the boundaries of mapping, to include virtual worlds and real world overlays with new sports, games, entertainment, and arts emerging.

As the climate continues to change, disasters will increase that will be both natural and involve human conflicts. Positioning and GIS systems will be at the forefront of emergency response activity with a real-time view of where all key people and equipment are located and automatic services will be used to advise people who need not be experts. Many developing countries will see Whereness as a key technology and not-for-profit initiatives will be deploying it to help with aid and conflict management.

Many businesses that today are exploiting basic ICT will see Whereness as the means to fill a revenue gap as the former core businesses commoditize more rapidly than predicted. A few organizations (perhaps only two) will compete to offer a comprehensive Whereness support service globally. It will, hopefully, be fully Semantic Web compliant and conform to a set of open standards that will be agreed as the Whereness ontology,

Most people and organizations will have accounts with one or more Whereness providers, the most basic of which will be free to use (but will be paid for by a mix of advertising, taxation, and revenue sharing with dependent services such as media sales, telephony and messaging, LBS, and ITS). The service will maintain a real-time and accurate log of position with an associated personal profile that will be under the direct control of the user (except in certain countries where government regulations will limit control or where the services will be managed through the government).

More advanced value-added services will also be available to businesses and will take the place of many current enterprise systems concerning supply chain management, logistics, workforce management, fleet management, and access security. Special niche services will be offered to the emergency services and the military that will compliment dedicated systems.

Overall it can be expected that Whereness may well follow the same commercial exploitation routes as the Internet and cellular radio. Although probably not quite as large in overall economic impact, Whereness may become a true utility and so ubiquitous it is taken for granted.

9.6 A Whereness "To-do" List for Today

This book is visionary, but visions are of little value without action. There is currently a wide gulf between what is possible and what is available today. Clearly, some concrete exploitation route maps are needed.

There are two key centralized services needed before Whereness can really start: first, commitment to the central Whereness service provision, and second, commitment to the management of the ubiquitous mapping content. The Whereness services are likely to be operated by large ICT organizations with global reach, whereas the mapping may be a more collaborative initiative, perhaps including aspects of OpenStreetMap and Wikipedia.

Whereness service provision will need to focus on system convergence, in particular taking feeds from all communications network providers and other organizations operating sensing, ticketing, surveillance, smart building systems, and so forth. GNSS service provision will also need to be included. Galileo services could, for example, be a stand-alone service and as part of a converged Whereness service bundle. Close cooperation will be needed with device manufacturers and these will be increasingly adding new wireless and sensor technology to their platforms.

The general public will be more demanding of Web services based on Whereness. Maps and images will be expected to be real time and be integrated with cameras and space imaging. They will be increasingly in a 3D format, appearing as scene views rather than a normal flat map view. The 3D images will include the inside of some buildings and also extend into virtual worlds. Some images may be video streams or animations based on the 3D models. Tools are therefore needed to create automatically all the visual components. Research work will continue and as the Semantic Web approach starts to automate the information behind Whereness, the Web will show dynamic models of the real world, with actual images rendered on the surfaces of the models with close associations with virtual worlds. For example, if I am represented by an avatar in a virtual world perhaps I would choose to be represented by the same character in the real world model. Perhaps in my home there would be a virtual portal to provide a means to cross between the real world and my virtual world. My friend's portals would all lead to the same virtual world where we might be meeting online and engaging in social activity.

Governments and large organizations can reap great rewards from Whereness, but currently the main barrier to Whereness is a lack of trust. Time and again when positioning is debated, people raise concerns about privacy and trust. The only way to deal with these issues is to be more open and allow users to view and control for themselves any information collected, stored, and shared if it involves them or things with which they have an interest. If organized optimally, Whereness can be a benefit to both users and organizations, but mutual trust and symmetry in control is needed first.

Rather than conclude on the rather negative Big Brother issue, perhaps it would be better to think of Whereness as part of something magic. It will enrich lives and extend the sense of self and the social group; it will lead to new fun and games and also could help to save the planet.

References

[1] Steventon, A., Wright, S, *Intelligent Spaces The Application of Pervasive ICT*, London, Springer-Verlag, 2006.

[2] Dennis, R, Wisely, D. "Mobility and Convergence," *BT Technology Journal*, Vol. 25, No. 2, Apr. 2007.

[3] Harle, R., Hopper, A., "Cluster Tagging: Robust Fiducial Tracking for Smart Environments," Location-and Context-Awareness Second International Workshop, LoCA, Dublin, Ireland, May 2006, pp. 14-29.

[4] Berners-Lee, T., Fischetti, M., *Weaving the Web: The Original Design and Ultimate Destiny of the World Wide Web by Its Inventor*, San Francisco: HarperSanFrancisco, 1999.

[5] WC3, "Technical Reports and Publications," http://www.w3.org/TR/#Recommendations, Jan. 2008.

[6] Noy, N.F., McGuinness, D.L., "Ontology Development 101: A Guide to Creating Your First Ontology," *Stanford Knowledge Systems Laboratory Technical Report KSL-01-05 and Stanford Medical Informatics Technical Report SMI-2001-0880*, Mar. 2001.

[7] Lacy, L.W., *OWL: Representing Information Using the Web Ontology Language*, Victoria BC, Canada:Trafford, 2005.

[8] Stahl, C.,Haupert, J., "Taking Location Modeling to New Levels: A Map Modeling Toolkit for Intelligent Environments," *Location-and Context-Awareness Second International Workshop, LoCA,* Dublin, Ireland, May 2006, pp. 74-85.

[9] Powers, S., *Practical RDF*, Sebastopol, CA: O'Reilly, Jul. 2003.

[10] Mullins, J., "Uncharted Territory," *New Scientist*, 31 May 2003, pp. 38-42.

[11] Schindler, G., Metzger, C., Starner, T., "A Wearable Interface for Topological Mapping and Localization in Indoor Environments," *Location-and Context-Awareness Second International Workshop, LoCA*, Dublin, Ireland, May 2006, pp. 64-73.

Epilogue

Digital Map Basics

Figure E1 shows the main ways to handle information map information (or indeed most other types of computer graphics). On the left, an image may start as a true picture taken by an analog camera using a film (of a light-sensitive chemical emulsion) or as an image drawn or painted by hand. Many maps still exist in this form so the first stage is to digitize then with an electronic scanner or digital camera. Most imaging today uses digital sensors that perform this digitization step inherently; the result is identical and is shown in the center.

A matrix that is usually rectangular quantizes and digitizes the image. By this we mean that the levels of light (or whatever energy is being captured) are placed into discrete levels and then the levels turned into numbers. In this simple example the simplest case has been chosen of two levels that are then encoded as 16-bit binary numbers. Usually, each dot (known as a pixel) would have perhaps 256 levels producing a grayscale leading to an eight times increase in information.

A further complication would be capturing a color image where each pixel would be in fact a triad of three pixels for red, green, and blue colors (from which any color can be created due to the nature of human eyes, which have similar sensors). Thus we now have a 24-fold increase of information per pixel. For a good resolution of a large image it can be appreciated why cameras are sold with megapixel resolution!

To reduce this huge information overload two approaches are taken. The first is to simply reduce unwanted redundancy (e.g., large white areas) using an encoding algorithm such as defined by the Joint Picture Experts Group (JPEG). A second approach is more powerful and difficult but turns the image into a parametric form using vectors (a quantity having both magnitude and direction). The right of Figure E1 shows how the matrix image (usually called a raster image) has been vastly simplified and, more importantly, placed in a format that can be manipulated using coordinate geometry mathematics in a computer. Each point or node has a coordinate pair (x,y) and the diagram shows how the vectors are made from pairs of points. During the algorithm to convert from raster to vector format it is normal to filter out unwanted speckles or "noise."

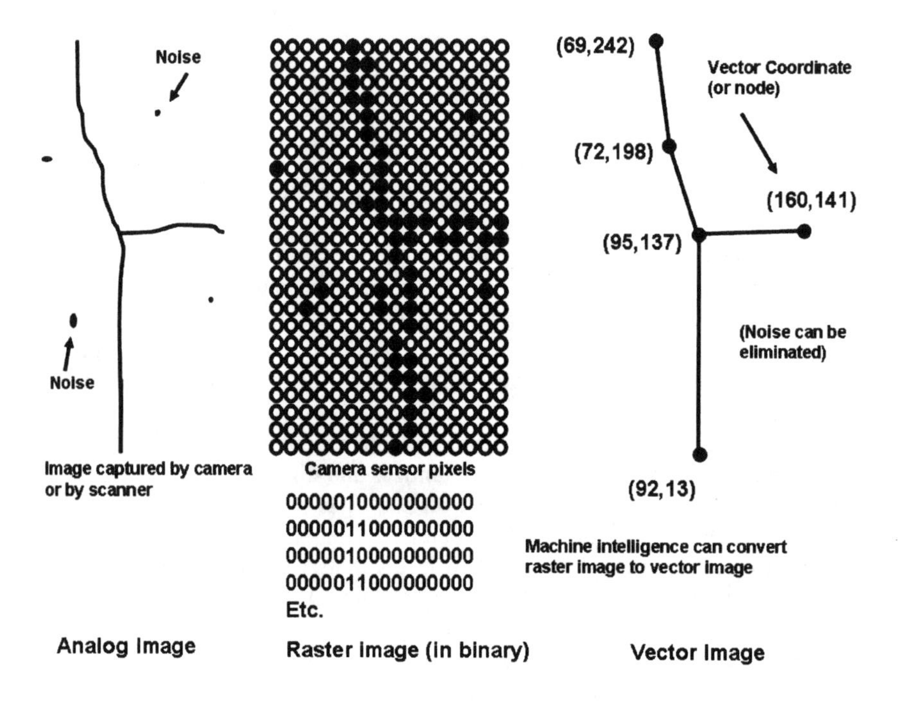

Figure E1 Graphical representation of the information about a simple map fragment.

Manipulating Vectors

The vectors can be linked together into graphs or graphical objects that can behave as individual entities. Thus it is easy to move one object (by varying the coordinates as shown above) while leaving the others in place.

The following list gives some examples of simple manipulations that could be performed on the fragment of the map:

- Pan right: add a constant to all x coordinates;
- Pan left: subtract a constant from all x coordinates;
- Pan up: add a constant to all y coordinates;
- Pan down: subtract a constant from all y coordinates;
- Zoom in: multiply all coordinates by a constant >1;
- Zoom out: multiply all coordinates by a constant <1.

Normally, when the graphic is required to be viewed, a raster matrix image is recreated since that is what most visual display units require. Originally, however, the cathode ray tube (as used in the first radars and still used in basic oscilloscopes) displayed vector graphics directly and the same principle is used today in laser light shows.

We can also start to think about three dimensions, where each node has three numbers (x,y,z) and where we use the term voxel rather than pixel (i.e. volume-pixel) for the sampled space. The vector geometry becomes a little more complicated, but rendering a 3D vector graphics in two dimensions is a heavy computer task which is why computer games require very powerful dedicated graphics chips.

In the example an arbitrary origin for the graphing has been chosen. As explained in Chapter 1, all maps have a datum or reference frame and we also know that there are both relative and absolute positioning systems, so is there a standard datum? Fundamentally, the universe has no reference frame datum since time and space are relative. Although relativity can usually be ignored in day to day applications, the exacting timing and speeds of spacecraft, for example GPS satellites, have to take it into account. The usual approach is to define a convenient datum and then to treat it as absolute (but relative to other absolutes used in other systems). The most common datum is the World Geodetic System's 1984 definition (WGS-84) which defined the earth as an ellipsoid (i.e. a solid ellipse slightly flattened at the poles) with a point centre.

To manipulate vectors centered on the Earth requires a polar coordinate system so care is taken to convert to and fro between Cartesian coordinate 3D vectors into polar coordinate versions. Mathematically, we are so far only dealing with Euclidean space (where the shortest distance between two points is a straight line) but other "spaces" are also theoretically possible (and like relativity, something only to trouble physicists).

When vectors need to be manipulated in complex ways such as distorting images to form a map projection or rotating a 3D solid image so a hidden area can be viewed, trigonometry is needed. A simple example for Figure E2 would be to rotate the vectors through $\theta°$ rotating about any point (a, b). See Figure E3. For each node then the following steps are needed to current the coordinates (x, y) to the new rotated coordinates (X, Y).

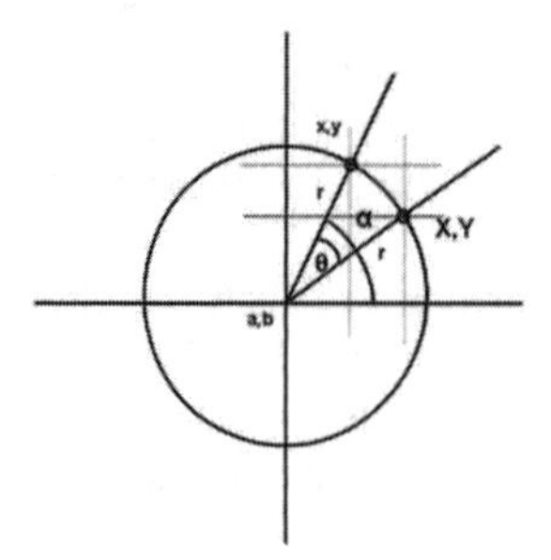

Figure E2 Rotation of a vector node.

Assume that a circle of origin (a, b) passes through (x, y) to the rotated coordinates (X, Y). It has a radius r.

$$r = \sqrt{(x-a)^2 + (y-b)^2}$$

The angle (from the x-axis) of r passing through (x, y) is $\alpha = \sin^{-1}(y/r)$
The angle (from x-axis) of r passing through (X,Y) is $\alpha - \theta$

Then $\begin{aligned} Y &= r.\sin(\alpha - \theta) \\ X &= r.\cos(\alpha - \theta) \end{aligned}$

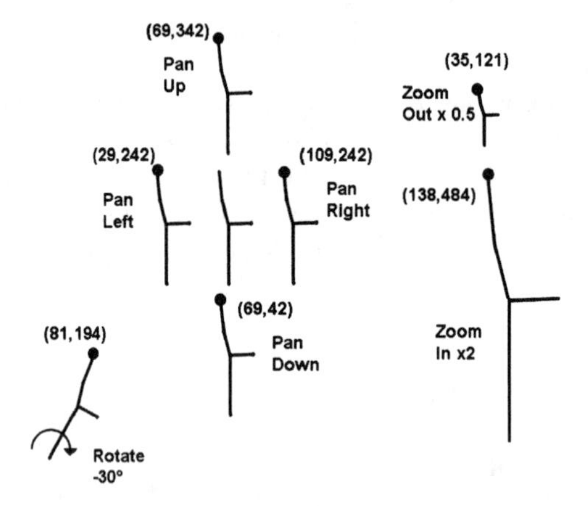

Figure E3 Manipulating vectors using coordinate geometry.

Physical Errors

All physical quantities are subject to error and although statistical techniques can reduce errors they cannot be eliminated. The error will be a probability distribution. To simplify processing we can set a reasonable probability level (say 90% likelihood) and thus reduce the range of values to a simple ± 5%.

So it is important to treat all quantities as a value and an error or alternatively as two values that represent the upper and lower bounds. This is helpful when considering whether (or not) an object's position is within a geo-fenced area or not. For example in Figure E1, all the coordinates of the nodes would be doubled so that each node would be the square of four nodes and a vector would be a rectangle joining the two nodal squares (see Figure E4). Also, the position of the object would not be one point but four points (the bounds of its error). A simple set of joint probabilities can then be calculated to determine if the object is present.

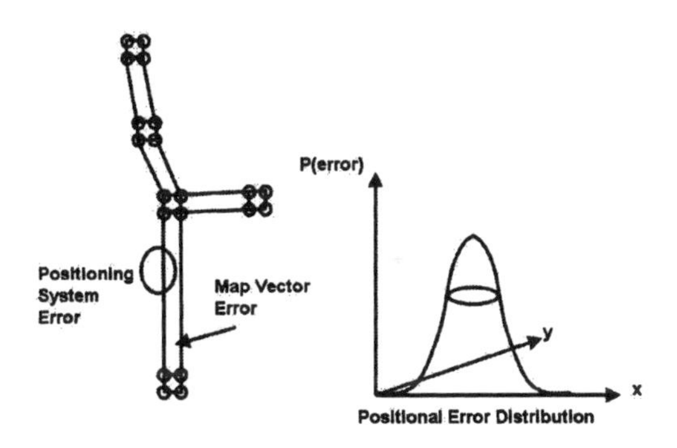

Figure E4 Map Vector Errors and Positioning Error

To recap, for each node in 3D space it is therefore possible to define 6 numbers.

Appendix—A Taxonomy of Positioning[1]

Physical Location

Physical positional information would include a set of coordinates, e.g., for a GPS receiver it might be latitude, longitude and height, or map references according to a standard grid such as a Ordnance Survey Grid Reference (OSGR). In human terms the information would be numerical.

Symbolic Location

Symbolic information is more abstract and would include the name of a room where a target is located or the area entered as defined as by a geo-fence. In human terms the symbolic approach is likely to be language-based or pictorial; when used by computers it is more concerned with associations between objects.

Reference Frame

Spatial systems have frames of reference that objects within the frame can use to share location information. It is sometimes useful to consider absolute and relative frames of reference but these are not innate qualities and depend upon what is known and what is being considered.

Reference Frame (Absolute)

Absolute location systems use a shared frame for all objects within the scope of the system e.g., GPS where the frame is based on Earth.

Reference Frame (Relative)

In relative location systems each object uses its own frame, e.g., a basic ship's radar system where the frame is centered on the rotating antenna and is displayed relative to the center of the rotating display.

[1] After Hightower, see Section 1.4.3.

Location Sensing Techniques

A number of sensing techniques can be used either singly or in combinations and some may be subdivided further.

Triangulation

Triangulation involves the use of geometry, measuring the angles and sides of triangles and thus calculating by trigonometry the coordinates of a required location from the position of two other vertices.

Triangulation by Lateration

Lateration involves repeated distance measurements. A 2D fix requires three distance measurements and a 3D fix four distance measurements. There are three general approaches:

Lateration by Direct Measurement

Direct linear measurements (e.g., using tape measure) would normally be difficult to obtain automatically. A wheeled vehicle could, however, use an odometer if traveling in a straight line.

Lateration by Time of Flight

Distance can be determined from accurate time measurement of direct signals traveling between two points (assuming a constant known velocity) and rejecting any reflections or multipath signals—apart from the special case of where a round trip measurement is made, e.g., in a conventional pulsed radar system where signals are bounced of a target.

Lateration by Attenuation

In free space conditions, radio signal attenuation falls proportional to the square of the distance between transmitter and receiver. Most urban and indoor environments are, however, dominated by multipath signals and subjected to additional statistical variations if one or the other (or both) stations are mobile. In these situations simple average signal strength can be a very misleading measurement.

Triangulation by Angulation

The repeated measurement of angle or bearing and often combined with lateration. For example, if the distance between two points is measured together with the two bearings to an unknown point, the coordinates of the unknown point can be calculated. For three dimensional location including height, an additional azimuth angle is needed.

Proximity

Proximity of an object to a sensor can be determined either by direct contact or by a nearness measurement which has a threshold set at the limit of contact. Clearly, there is some commonality between this approach and the lateration methods discussed above in particular in the wireless case where the presence of a wireless signal (above the detection threshold) signals "contact". Proximity thus gives symbolic information rather than a physical measurement.

Physical Contact

Large scale systems include active floors using pressure switches to determine foot fall, door switches, window switches, lock and key systems, body pressure switches on furniture.

Electrical Contact

The interconnection of electrical apparatus can signal proximity. Dedicated proximity systems such as proprietary contact tags (e.g., i-button) and electronic ticket machines are used directly. Indirectly many other devices can be used to signal proximity including wired telephones, computers on LANs, PDAs in docking stations, and smart cards in smart card readers. Some vehicle alarms use more subtle approaches—detecting activity by changes in electrical power consumption and thus signaling the (symbolic) location of an intruder.

Wireless Contact

This is an area of great importance to position systems since cheap, portable and low power sensors and transmitters can be used. A wide range of infra-red, radio, ultrasonic systems exist. In addition passive detection can be used e.g. PIR sensors for body heat, and capacitive sensors for body proximity. Magnetic sensing is used in robotic systems. The RFID industry used quasi-passive EM field powered tags for supply chain management.

Inertial Navigation

Inertial systems rely only on internal sensors such as gyroscopes and accelerometers and have no need (once calibrated from an external reference) for any outside signals. They are used for military and aerospace applications since they are both reliable and immune from jamming. They disadvantage is that long term drift can cause errors to accumulate and the systems are expensive and cumbersome. Often combined with map matching / terrain following systems.

Mechanical

Vehicle navigation systems can use wheel sensors (often ABS brake sensor pulses) to provide both distance traveled and turning information (from differential signals from both wheels on one axel). A crude approximation can be used for pedestrians where the pedometer measures foot step numbers usually by means of a small mechanical pendulum located near the hip. Average foot step length can then yield an estimate of distance traveled.

Magnetic Compass

A crude approximation of heading can be found by using a flux-gate electronic magnetic compass or a wireless compass based on radio broadcasts.

Processing (Centralized)

Most positioning systems apart from GPS and inertial systems employ centralized computation of position. The advantages are shared processing costs and the ability to track / interrogate directly and combine with other information sources.

Processing (Autonomous and Localized)

GPS and inertial systems use local and autonomous processing. The advantages are that there is no limit to the number of devices that can be positioned, and privacy is under the direct control of the person/object being tracked.

Processing (Distributed and Federated)

A distributed and federated approach to positioning (which is a hybrid that combines centralized and localized positioning) has the advantage of mobile devices sharing location information using peer-to-peer communications thus improving accuracy and lowering costs of fixed infrastructure.

Errors

The quality of any positioning system might be given, as for example, 10m approximately 95% of the time. Various estimation techniques are used to improve the quality of readings by combining measurement (e.g., using Kalman filters) and by combining the outputs from different types of sensors (using fusion techniques).

Error (Accuracy)

Accuracy is the distance between the given position from the true position (e.g., 10m typical for GPS)

Error (Precision)

Precision is the proportion of readings where the accuracy that has been stated holds true (in the GPS case perhaps 95%).

Sensor Fusion

Simultaneous use of different location sensing systems can be advantageous by increasing accuracy and precision beyond that which could be obtained using single systems. Hierarchical and overlapping data sets can aggregate properties and combine differing error distributions using Markov and Bayesian statistical techniques.

Adaptive fidelity

Adaptive fidelity is a technique to change precision in response to system factors (e.g., battery power failing).

Broadcast

Broadcast systems such as GPS have no scale problem since they rely on a global set of broadcasts that have virtually no limits to the number of receivers that can receive their signals. In contrast, many other position systems (e.g., active badges) use a fixed infrastructure and a centralized approach to positioning that will always have an upper limit on the number of terminals it can support.

Point to Point

Point to point communications use a direct 2 way path that is used, for example, to interrogate an active tag and capture the information about its response—which could be its identity (symbolic) or its (physical) range or both.

Latency

Latency is the time needed to get a position or track from the target or system. If the time is too long the information may become irrelevant to the application—for example, in a fast moving vehicle, the destination may have been passed before the information is known.

Identification

Most positioning systems have subsystems to identify the user of the system or the system being used or both. Recognition technology can recognize shapes and electronic identity can be communicated. Ideally a Global Unique Identification (GUID) would be used.

List of Acronyms

3D	Three Dimensions of Space
4D	Four Dimensions (3D space plus time)
2.5G	Cellular Radio Generation between 3G and 2G
3G	Third Generation Cellular Radio System
4G	Fourth Generation Cellular Radio System
ABS	Anti-Lock B
AC	Alternating Current
ADF	Automatic Direction Finder
AGPS	Assisted GPS
AI	Artificial Intelligence
AJAX	Asymmetric Javascript and XML
AoA	Angle of Arrival
AOL	Corporation Name
API	Application Programming Interface
ARPU	Average Revenue per User
ASIC	Application Specific Integrated Circuit
BT	Corporation Name
CCD	Charge Coupled Device
CCTV	Closed Circuit Television
CDMA	Code Division Multiple Access
Cell ID	Cell Identification
CLI	Calling Line Identity
CMOS	Complementary Metal Oxide Semiconductor
CPGPS	Carrier Phase (enhanced) GPS
DAB	Digital Audio Broadcasting
DF	Direction Finding
DGPS	Differential GPS
DNE	Digital Networked Economy
DRG	Dynamic Route Guidance
DRM	Digital Rights Management
DSRC	Dedicated Short Range Communications
DSP	Digital Signal Processor
E911	Emergency Call Regulation in USA
ECGI	Enhanced Cell Global Identity
EGNOS	European Geostationary Navigation Overlay Service
EOTD	Enhanced Observed Time Difference
ESA	European Space Agency
ESRI	Corporation Name
E.U.	European Union

FAA	Federal Aviation Authority (of US)
FOAF	Friend of a Friend
FM	Frequency Modulation
GIS	Geographical Information system
GLONASS	Global Navigation Satellite System (of Russia)
GML	Geographical Markup Language
GNSS	Global Navigational Satellite System
GPRS	General Packet Radio Service
GPS	Global Positioning System (of US)
GSM	Global System of Mobile Communications
HAP	High Altitude Platform
HF	High Frequency
HMI	Human Machine Interface
HTTP	Hypertext Transfer Protocol
HTML	Hypertext Markup Language
HUD	Head Up Display
ICT	Information and Communications Technology
ID	Identity
IFF	Identification Friend or Foe
IP	Internet Protocol
IR	Infra-Red
ISM	Industrial Scientific and Medical (radio band)
ISP	Internet Service Provider
ITS	Intelligent Transportation Systems
IVHS	Intelligent Vehicle Highway Systems
KML	Keyhole Markup Language
LAN	Local Area Network
LBS	Location Based Services
LED	Light Emitting Diode
LEO	Low Earth Orbit
LF	Low Frequency
LIDAR	Light Detection and Ranging
LMU	Location Monitoring Unit
LORAN	Long Range Navigation
MEMS	Micro-Electro-Mechanical System
MIMO	Multiple Input Multiple Output
MIT	Massachusetts Institute of Technology
NFC	Near Field Communications
NMEA	National Maritime Electronics Association
OGC	Open Geospatial Consortium
OS	Ordnance Survey (corporation name)
OSM	Open Street Map
OWL	Web Ontology Language
PC	Personal Computer

PDA	Personal Digital Assistant
PIR	Passive Infra-Red
PMR	Private Mobile Radio
PoI	Point of Interest
QoS	Quality of Service
RADAR	Radio Direction and Ranging
RDF	Resource Description Framework
RDS	Radio Data System
RDS-TMC	RDS Traffic Message Channel
RFID	Radio Frequency Identification
RSSI	Received Signal Strength Identification
SLAM	Simultaneous Location and Mapping
SMS	Short Message Service (of GSM)
SML	Sensor Markup Language
S/N	Signal to Noise Ratio
SONAR	Sound Navigation and Ranging
TA	Timing Advance
TCP/IP	Transfer Control Protocol of IP
TDOA	Time Difference of Arrival
TMC	Traffic Message Channel
TOID	Topographic Identifier
UHF	Ultra High Frequency
URI	Universal Resource Identifier
U.K.	United Kingdom
URL	Universal Resource Locator
U.S.	United States (of America)
USGS	United States Geographical Survey
U-TDOA	Uplink TDOA
UWB	Ultra Wide Band
VHF	Very High Frequency
VOIP	Voice over IP
VOR	VHF Omni-directional Range
VR	Virtual Reality
W3C	World Wide Web Consortium
WDM	Wavelength Division Multiplexing
WGS-84	World Geodetic Standard 1984
WAAS	Wide Area Augmentation Service
WLAN	Wireless LAN
WW2	World War 2
WWW	World Wide Web
XML	Extensible Markup Language